R2701732   20.95

Micrographics
A User's Manual

MICROGRAPHICS.
A USER'S MANUAL

JOSEPH L. KISH, JR.

Vice-President
FENVESSY ASSOCIATES, INC.

A WILEY-INTERSCIENCE PUBLICATION

JOHN WILEY & SONS
New York • Chichester • Brisbane • Toronto

Library of Congress Cataloging in Publication Data

Kish, Joseph L, Jr.
   Micrographics.

   "A Wiley-Interscience publication."
   Includes indexes.
   1.   Micrographics—Handbooks, manuals, etc.
I.   Title.
Z265.K53   686.4'3   80-16798
ISBN 0-471-05524-7

Printed in the United States of America

10 9 8 7 6 5 4 3 2 1

To My Parents

PREFACE

Increasingly, business, governmental, educational, and nonprofit organizations are addressing the question of how to improve the productivity of their office personnel. Faced with escalating salaries and space costs, these organizations are subjecting their office operations to the same close scrutiny and cost containment that they have applied to their manufacturing and distribution operations for decades. Cost-conscious managements have realized that any improvements achieved in office productivity yield savings that go directly to the bottom line of their Profit and Loss Statement.

In such an environment, managers are showing an active interest in other methods of creating, disseminating, filing, and retrieving correspondence, forms, reports, publications, and the various other records that their organizations receive and generate. They are seeking methods of recording business activities and communicating that are more efficient and economical than the paper (hard copy) formats that have been in use since Biblical times.

This book is designed to assist in that effort by providing information that is useful to both the manager and the information systems technician with respect to one primary alternative to paper—*micrographics*. To the manager, it provides an understanding of the micrographics process, its advantages, limitations, and costs. It addresses such salient, business-oriented questions as: Can I legally use microfilm in my business? Will the micrographics systems and hardware I install today be made obsolete by the "office of the future"? What type of people must I hire to install and maintain the micrographics-based systems?

For the systems technician, who must evaluate the feasibility of micrographics and plan for its cost-effective implementation, this book is intended as a continuing source of both practical and technical information. It provides guidance in answering such questions as: What possible applications exist in our business for both source document micrographics and computer-output-microfilm (COM)? How can I accurately determine the cost and operational feasibility of a micrographics-based system in a given application? What indexing techniques may I use to facilitate rapid retrieval of a given microimage? Should I consider purchasing used hardware to reduce my capital investment?

Micrographics—A User's Manual has been written from the viewpoint of one who is *not* a micrographics specialist. It employs terminology that is understandable to all business. The methods of presentation, the illustrations, and the applications have all been prepared with one purpose in mind: to provide the reader with a clear, concise, understanding of micrographics and an appreciation of how this technology can be used in his or her own organization to reduce operating costs while improving information creation, dissemination, storage, and retrieval.

I want to express my gratitude to a number of persons for their contribution to the completion of this book: to Grace Kish and Paul Selzam for the preparation of illustrations; to four busy men who took time to review the manuscript and suggest improvements—Lou Lachter of 3M, Bill Wnek of E. R. Squibb & Sons, Belden Menkus of Menkus Associates, and Mario Crespo of Avis; to Frances Kish for her assistance in typing and preparing the format; and to Gerry Papke, Linda Grady-Troia, Linda Dugan, and their staffs for a fine editing and production job.

Finally, for permission to adapt material I have previously published, a special thanks is due to *Administrative Management*, a publication of Geyer-McAllister Co., as well as to *Business Graphics* and *Graphic Arts Monthly*, each a Dun-Donnelly publication.

JOSEPH L. KISH, JR.

Westfield, New Jersey
July 1980

CONTENTS

Micrographics
A User's Manual

An Introduction to Micrographics

Micrographics, the technology involved in the creation and use of microfilm images, is today, nearly 125 years after its initial development, truly coming of age. Business, government, educational and non-profit organizations of every type and size are utilizing microfilm in an effort to economically and efficiently organize, store, retrieve, reproduce, distribute, and display information formerly maintained as paper, or in such computer-processable formats as magnetic disks, magnetic tapes, and punched paper tapes.

How can this tremendous interest in microfilm be explained? How can it be that microfilm, which just 20 years ago was utilized by most organizations solely as a means of retaining old, infrequently referenced records in a less space-consuming format than paper, has now evolved into a viable information management concept for all types of records, regardless of their format or frequency of reference. To answer these questions, one must first examine the basic problems facing organizations as the result of the proliferation of information and records they receive, generate, maintain, and reference.

THE "INFORMATION AGE"

We currently are living in "The Information Age." Ninety percent of all the mathematicians, engineers, scientists, and technicians who have ever lived are alive and employed today. Together with the ever-increasing white collar work force (which for the first time in modern history is approaching the numbers of production and distribution workers), they are generating tremendous volumes of paperwork annually.

To appreciate just how vast these volumes of paperwork are, consider that according to a recent Xerox Corporation study:

- American businesses currently maintain more than 324 billion documents in their files.
- Each year, this volume grows by approximately 22% (about 72 billion documents).
- Based upon this average annual growth, the total volume of records maintained by American business organizations will double every five years.

The typical organization's records comprise a virtual warehouse of information upon which its management may base its policy, financial, and operating decisions. The most complete records in the world, however, must be organized into a readily and accurately retrievable format to be usable and valuable. Just how to accomplish this organization and formatting is the problem. The computer, although able to analyze data, perform mathematical calculations, control complex production operations, and perform simple, logic-based decisionmaking, is far from being an economical, efficient means of storing and retrieving data that is of medium to long-term reference value.

The vast majority of records, due to their format and reference requirements, cannot be economically reduced to a computer-processible form or be stored in computer memory. Such records must remain in a human-readable format for the present, at least. Until recently, the necessity of retaining records in a human-readable format meant just one thing—hard copy (paper) storage and retrieval, and all the problems inherent in that method, such as banks of files, quantities of filing supplies and floor space, the need for a variety of controls and procedures to police the return of borrowed records, the difficulties posed by lost, misfiled, damaged, and pilfered records, and so forth.

Microfilm held much promise for solving many of these problems, but for years, the state-of-the-art left so much to be desired, and the acceptability of the microfilm in courts of law and in legal procedings was so left to the discretion of the trial judges, that microfilm was looked at purely as a means of storing and retrieving archival and other infrequently referenced records. Microfilm was basically relegated to the basement storage area—paper was the primary means of storing *active* records and information.

Ten years ago, based upon the author's experience, many business executives, while allowing that microfilm was an economical way of storing rarely referenced records and information, would follow up by stating that it was inappropriate for use in *active* information storage and retrieval applications because it was "too complex," "too hard to index," "too expensive in terms of hardware investments,'' ''subject to too many quality control problems,'' and ''unacceptable to too many legal and regulatory bodies both in the United States and abroad.''

Faced with the problem of overcoming these and other user objections, or never rising from basement storage applications, the various members of the microfilm industry set to serious efforts to improve their technology, quality standards, indexing techniques, equipment performance, and reliability and to simplify the entire microfilm procedure in general. It was a long, difficult process, but they have been successful. Every major objection stated earlier, and virtually all of the minor objections (other than personal preference) have been overcome. The state-of-the-art has advanced to the point where equipment is easy to operate, relatively trouble-free, and is available for purchase or lease at prices geared to nearly every user's means. New technologies, such as high reduction microfiche and micropublishing have made microfilm a far less expensive media than hard copy. Improved quality controls and standards, coupled with better understanding of what must be done to assure long-term preservation of microfilm records, have calmed many doubts so far as the permanency of microimages is concerned. The development of improved indexing methods have made the location of individual images just as fast and economical, if not more so, than with hard copy reference. What is probably most important, microfilm has gained widespread acceptance as a substitute for hard copy records, both in the United States and in foreign countries.

Today, American businesses are scrutinizing their office operations, seeking new, more efficient, more economical ways of operating. The ever-increasing costs of office space, equipment, and personnel, coupled with the high costs of borrowed money, the uncertain economy, and the vast processing capabilities of the computer, has convinced many cost-conscious managements that they cannot afford to ignore their information storage and retrieval operations, if they are to provide efficient, responsive, cost-effective service. This re-examination and re-evaluation is resulting in a tremendous growth in the use of microfilm for active information storage and retrieval (filing and finding) purposes. There is no major area of administration in which microfilm is not now used as an active system.

ADVANTAGES OF MICROFILM

Microfilm has the following advantage over conventional hard copy (paper) records:

• **Requires Less Floor Space and Filing Equipment.** The use of microfilm will significantly reduce the floor space and filing equipment required for document storage. To illustrate: a four drawer letter-size file cabinet consumes 6 square feet of floor space and houses about 10,000 pieces of paper. The same volume of records could be stored on 4 rolls of microfilm that would occupy a total area of 4 cubic inches—a space reduction of approximately 98%.

• **Low-Cost Reproducibility.** Microfilm may be duplicated for distribution or security purposes at a very nominal cost. A 100-foot roll of microfilm containing approximately 2,400 images may be duplicated for about $6.50; a microfiche containing 98 microimages for approximately 50¢.

• **Relative Permanence.** If stored in a suitable environment in which the temperature and relative humidity are controlled, and in which no acidic gases are present, and if reasonable precautions are taken to minimize the incidence of mishandling, microfilm will last almost indefinitely. With microfilm, one need not worry about dog-eared, torn, faded, or improperly repaired documents.

• **File Integrity.** The likelihood that a given record will be misfiled is greatly reduced when microfilm is used.

• **Assures Low-Cost Document Accessibility.** Microfilm enables more data to be furnished users at less cost than any other medium. Thus, each user can receive a complete working file of data—an action that would be too costly, in terms of duplicating and storage, with any other medium.

• **Reduces Mailing Costs.** Microfilm is less expensive to mail than is paper. An aperture card containing an engineering drawing can be mailed for 50 to 25% of the cost of the original paper drawing. A microfiche file can be mailed for about 2% the cost of the paper equivalents.

• **Low Maintenance Costs.** Once the initial costs of the microfilm conversion have been incurred, microfilmed records cost virtually nothing to maintain, in terms of filing equipment and floor space. Paper records, on the other hand, will typically cost about 1¢ per page annually, to maintain in the typical office's files.

LIMITATIONS AND DRAWBACKS OF MICROFILM

There are a number of limitations and drawbacks that one must be aware of and weigh against the foregoing cost and operational advantages, before any decision is made to substitute microfilm for paper records in a given application.

• **A Controlled Storage Environment is Essential.** Microfilm is subject to certain types of deterioration that paper records are not. Since the silver halide microfilm (the type typically used to produce negative microfilm formats), utilizes a gelatin emulsion—the same substance scientists use to grow cultures—it is extremely vulnerable to deterioration due to excessive humidity and temperature variations as well as the presence of such acidic gases as sulfur dioxide. Microfilm must therefore be stored in an environmentally controlled area in which the relative humidity remains constant between 35 and 40%; temperatures between 60 and 72° F, and the air is either free of acidic gases or is filtered to remove them. Microfilm also is subject to deterioration due to improper processing that leaves too much residual sodium thiosulfate penthydrate (hypo) on the emulsion of the microfilm.

• **Inability to Annotate.**   Microfilmed records cannot be annotated. Therefore, recipients of microfilm-based reports cannot make notations in the margins, nor can they enter data on the report as a means of updating it pending the next issue, nor can they correct mistakes, and so forth.

• **Difficult to Compare.**   It is difficult to compare two or more microfilm-based reports, forms, or other documents simultaneously. Usually unless the user works with two microfilm readers that are placed side by side, he must resort to the creation of a reproduction of one of the microimages.

• **Unable to Distinguish Colors.**   Microfilm is panchromatic, and will not reproduce colors. Therefore, if records to be converted to microfilm are color-coded or if they contain significant entries in color, either an alternative method of indicating the colors must be used, or color microfilm, a much more expensive alternative, must be employed.

• **Need for Specialized Reference and Use Equipments.**   Due to its miniaturized size, microfilm requires special-purpose equipment to enlarge it to a size that can be read by the naked eye or that can be reproduced in a paper format for subsequence reference.

• **Legal and Regulatory Restrictions.**   Various federal, state, and municipal governmental agencies have promulgated numerous, often contradictory regulations concerning the introduction of microfilm-based records as substitutes for the original hard copy documents. Depending upon the agency involved, these regulations may (for example) require the maintenance of the original hard copy for specified minimum periods of time, regardless of the availability of the microfilm copies; they may mandate the creation of a duplicate security copy of the microfilm; they may stipulate that the microfilm must meet certain film processing and image quality standards; and so on. The user's failure to comply with such regulations may result in the inadmissability of the microfilm records in courts of law and in regulatory proceedings.

• **Difficult to Browse.**   Many users find microfilm much more difficult to browse than either bound or individually filed hard copy records.

TODAY'S APPLICATIONS FOR MICROFILM

Four major uses are being made of microfilm today:

• As an active recordkeeping media.
• As a communications media.

- As a security media.
- To facilitate compliance with governmental and internal recordkeeping requirements.

Active Recordkeeping Media

- A stock brokerage firm was able to eliminate the need for additional office space for its research library by installing microfilm-based storage and retrieval systems.
- A consumer finance company installed a microfilm-based system for its loan files and was able to reduce personnel, equipment, supplies, and floor space costs by over $50,000 annually.
- A labor union converted its membership files from computer print-out to microfilm, and achieved significant reference efficiencies and economies.
- A construction company that installed a microfilm-based engineering recordkeeping system realized operating cost reductions of over $45,000 annually, achieved increased document security, and virtually eliminated the incident of lost, damaged, or misfiled drawings.

These typical "success stories" are among the reasons many organizations are evaluating the feasibility of microfilm as an *active* recordkeeping media.

The various impediments to the use of microfilm for any but the least referenced, least valuable of an organization's records have all been corrected. As noted previously, the state-of-the-art is such that microfilm is, if properly processed, formatted, indexed, and stored, as convenient a recordkeeping media as is hard copy. Records stored on microfilm are just as easy to retrieve as are the original hard copies. They are less subject to wear and tear. They are acceptable in courts of law and regulatory proceedings as are paper records (providing certain quality and indexing requirements are met). In many cases, the substitution of microfilm for hard copy records will lead to significant cost reductions and operating efficiencies.

The computer has played a major role in the emergence of microfilm as a viable active recordkeeping media, since it has made possible the development of improved, in-depth microfilm indexing techniques. The computer has the ability to interface with microfilm-based systems to prepare and update the indexes to the various microimages. This enables users to quickly and accurately pinpoint the *exact* microimage that they require to complete their reference and to locate the exact location of that microfilm image on the microfilm roll, microfiche, or whatever format is involved.

As the costs of floor space, filing equipment and supplies, and clerical and professional salaries continue to escalate, there will be increasing utilization of microfilm as an active recordkeeping media.

Communications Media

As organizations grow and diversify, expand their product lines or offer new services, open new markets and move into new geographic areas, the necessity of speedy and economical communications becomes increasingly important. Operating statistics and problems must flow upward from decentralized field and plant locations to central administration for decisionmaking and control purposes. Conversely, directives and policy formulations must be accurately and efficiently disseminated from headquarters to subordinate organizational levels. Microfilm, due to its ease and speed of creation and duplication, its image quality and efficient, economical mailing, storage, and retrieval, is ideally suited for both internal and external communication, and is finding increasing acceptance as a business communications media among cost-conscious administrators.

Microfilm-based business communications systems include a wide variety of record categories and numerous microfilm formats. The particular document dissemination and reference requirements will largely determine which format is the most applicable. These categories, and the microfilm formats in which they are typically prepared, are as follows:

- **Transaction Records.** Correspondence, invoices, purchase orders and the like are typically maintained in a roll, cartridge or cassette format.
- **Computer Output.** Statistics, forms, graphics, lists, and the like are maintained in a roll film or microfiche format.
- **Engineering and Technical Documentation.** Engineering drawings, tracings, bills of materials, specifications, and the like are maintained in a microfilm aperture card format.
- **Micropublishing.** Catalogs, reference materials, reports and other similar materials are maintained on microfiche.
- **Case Files.** Personnel files, project files, customer files, and the like are prepared in an updatable microfilm jacket format.

Until recently, many organizations have hesitated to utilize microfilm as a business communications media due to two primary factors:

- The difficulties involved in referencing the microfilm and making copies of individual microimages.
- Dissatisfaction with the quality of the microimage and the enlarged reproductions made from that image.

Micrographics technology has progressed to the point where both these objections have been overcome. Microfilm readers and reader-printers have reached a uniformly high level of performance and simplicity. Today, their operation may be quickly mastered. The micrographics optics, film, and operational practices have progressed to where the quality of the microimage itself and that of the enlarged reproduction that is displayed on the viewing screen of the reader or reader-printer, as well as the quality of the enlarged paper reproduction made from the microimage is of a quality that approximates that of the original record.

Microfilm provides an efficient, economical, effective means for any organization to reduce the costs of its business communications systems. In these days of rising costs and the need for current, usable information, microfilm could well prove to be a more convenient, more usable format for communication than hard copy-based computer print-out, engineering documentation, printed books, technical publications, and the like. Organizations of every size and type would be well advised to evaluate the feasibility of such microfilm-based communications systems.

Security Media

Many organizations utilize microfilm as their prime vital records protection media. Such organizations create an additional microfilm copy of those records that they have identified as containing information that is required for the reconstruction of the organization's essential operations as well as for the reconstruction of its legal and financial situation. This additional microfilm, commonly called the *security* or *vital records copy,* is then forwarded to a distant location for safekeeping, to be referenced only if the original record is lost or damaged. The concept behind the creation of a security microfilm copy and its storage in a different location from that of the original record is to preclude the possibility that both the original and the security copies will be destroyed by the same natural or manmade disaster.

Microfilm is ideally suited for use as a security media. Its compactness assures that maximum numbers of records will be stored in a minimum amount of space and equipment. Furthermore, microfilm will generally prove to be the fastest, most economical means of producing high-quality, duplicate copies of large numbers of records.

It should be noted that many regulatory agencies, such as the Internal Revenue Service and the Securities and Exchange Commission, stipulate that as one of the conditions for using microfilm in lieu of hard copy records, organizations subject to their jurisdiction must create a security copy of all microfilm records subject to the agency's audit, and maintain these microfilm duplicates in a site that is physically removed from the original microfilms.

Compliance with Governmental and Internal Recordkeeping Requirements

Every organization regardless of size or area of activity—from the corner candy store to the largest of the multinationals—is subject to the regulatory authority of a host of federal, state, and local governmental agencies. To comply with the recordkeeping requirements of such agencies, they all must maintain specific information relating to various aspects of their operations, and they must be able to retrieve such information upon demand by the regulatory agencies.

During the past two decades there have been major changes in governmental regulatory activity at all levels. As the result of new social and economic legislation and the refinement of computerized auditing techniques within, there has been a significant increase in the governmental regulation of businesses. The Equal Employment Opportunity Act, the Occupational Safety and Health Act, The Privacy Act, the Consumer Protection Act, revisions to the Social Security and to the Internal Revenue Laws all have resulted in businesses of every type being required to create and maintain information, unnecessary for internal use, to comply with the reporting and recordkeeping requirements of a plethora of governmental agencies. The cost of such compliance, which must be borne by the business alone, is fast becoming a significant part of operational expenses.

Consequently, most business are retaining large volumes of records solely to comply with regulatory agencies' reporting and recordkeeping requirements. In doing so they are incurring continuing, substantial expenditures for filing equipment and supplies, floor space, and personnel; these expenditures frequently come right off the bottom line of their Profit and Loss Statement.

Coupled with these legal and regulatory requirements is the organization's own reference needs. To provide the data required for the formulation of future decisions and plans, a variety of operational, financial, and other records must be retained for varying periods of time after their initial processing.

Therefore large volumes of records are required for long periods of time, which makes microfilm-based storage a distinct possibility. Many organizations are finding that microfilm will significantly reduce the costs involved in the long-term retention of those older records that are no longer required for continual, active reference, but that must be maintained to satisfy legal, regulatory, or internal reference needs.

In Chapter 5, the various legal, regulatory, operational, and cost ramifications of microfilm will be discussed in detail. Each of these must be carefully analyzed to assure that there is no prohibition or impedence to the use of microfilm to facilitate compliance with governmental and internal recordkeeping requirements.

TYPES OF MICROGRAPHIC SYSTEMS

There are two distinct types of micrographic systems:

 • **Source Document Microfilm.**   The microimages are prepared by photographing the actual paper records.

  • **Computer-Output-Microfilm (COM).**   A process in which the microimages are prepared from digitized data stored on disks or magnetic tapes, or transmitted directly from a computer's Central Processing Unit (CPU). No paper records are required in the COM process. The individual microimages are created by "writing" directly on the microfilm or by photographing the data as it is projected on the face of the COM Unit's Cathode Ray Tube (commonly referred to as a CRT).

These two processes, Source Document Microfilming and COM, are readily learned, and provide an organization with complete system capabilities. The Source Document Microfilming process enables written, typed, printed, and drawn materials, such as forms, correspondence, charts, graphs, and other similar hard copy records to be converted to a microfilm format. The COM process enables one to convert digitized, computer-processable data directly to a microfilm format, without having to incur the additional time, labor, equipment, and expense involved in creating and then microfilming a hard copy printout.

Basic Systems Flow—Source Document Micrographic Systems

The basic Source Document Micrographics system involves six steps:

- Document preparation.
- Filming of the records.
- Processing of the exposed microfilm.
- Inspection and formatting of the microimages.
- Creation of additional copies of the microimages for distribution and security purposes.
- Refencing the microimages.

Document Preparation

Prior to the actual microfilming, various document preparation must occur:

- All records to be microfilmed must be recalled from borrowers and returned to their proper positions in the file.
- Any torn documents must be repaired.
- Dirt, stains, and smudges must be removed from the face of the documents.
- All staples and paper clips must be removed.
- The records must be arranged so that they are all facing in the same direction.
- A Title Target, a Certificate of Authenticity Target, and any applicable indices and descriptive targets must be prepared and inserted in their proper locations among the records awaiting microfilming.

Microfilming

The actual microfilming of source documents is a standard photographic process, in which original paper records are photographed onto silver halide microfilm at considerably reduced sizes (e.g., a record filmed at a 16× reduction ratio is reduced to 1/256th of its original size). Depending upon the format and dimensions of the source document and the reference requirements for the resultant microfilm, one of three types of microfilm cameras will be utilized.

• **Rotary (or Flow) Camera.** Designed to microfilm source documents that are generally legal size or smaller (i.e., up to 8.5 × 14 inches) and are individual pieces of paper. Examples of such source documents would be correspondence, noncomputer-prepared business forms, and unbound reports. With the rotary camera, individual source documents are automatically photographed as they pass in front of the microfilm camera's lens. The output of the rotary camera is roll microfilm, typically in continuous 100-foot lengths.

• **Planetary Camera.** Designed for the *precision* microfilming of bound books and documents, as well as engineering drawings and tracings and those source documents that are too large to be fed into a rotary microfilm camera. During the actual microfilming, the record being filmed remains stationary and is filmed in a single exposure. As the result, distortion caused by such factors as the source document and microfilm moving at different rates of speed (which is often the cause of blurred images when the rotary microfilm camera is used), is not possible. The output of planetary camera microfilming is roll microfilm in either 16-, 35-, or 105-mm widths, typically in continuous 100-foot lengths.

• **Step-and-Repeat Camera.** A special-purpose planetary camera with a microfiche output. Microfiche is a microfilm format in which the various microimages are arranged on a sheet of negative microfilm, much in the same manner that the days of the month are arranged on the pages of a calendar—sequentially in rows and columns in a left-to-right, top-to-bottom order.

Microfilm Processing

After exposure, the silver halide microfilm is developed, using either a commercial microfilm processing laboratory or by the organization's own personnel on a microfilm processor. In the developing process, the exposed microfilm is passed through a series of chemical baths that develops the negative image and makes it impervious to subsequent exposure to light.

Inspection and Formatting

After developing, the negative microfilm is examined for image quality using a microfilm reader or, in some instances where minimum quality levels must be assured, using such specialized testing equipment as densitometers. The microfilm reader is capable of enlarging individual microimages to a size that can be read with the naked eye, and projecting this enlarged image on a ground glass screen for direct viewing. During the inspection, any illegible or otherwise unsatisfactorily-filmed microimages are noted, so that the corresponding source documents may later be remicrofilmed.

If the microimages are to be maintained in other than a roll format (e.g., inserted into microfilm jackets, or stripped-up to form microfiche), such formatting will then occur.

Duplication and Distribution

After the microimages have been converted to their final format, any additional copies required for distribution or document security purposes will be created using specialized microfilm duplication equipment. The resultant duplicate microfilm may be prepared as either silver halide microfilm or on diazo or vesicular films.

Reference

When it is necessary to reference the microfilm, it is merely inserted into a suitable microfilm reader or reader-printer (the latter being a microfilm reader that has the additional ability to produce enlarged paper reproductions of selected microimages). The desired microimage is then located and is projected on the reader or reader-printer's ground glass viewing screen. After examining the enlarged microimage on the viewing screen, if a paper reproduction is required, the referencer need only push the reader-printer's *PRINT* button and in 5 to 15 seconds, an enlarged paper reproduction will be produced, ready for use.

In Chapter 4 the production and use of source document microfilm will be examined in greater detail. The features and operation of the microfilm readers, reader-printers, microfilm processors, and the various duplicators will be examined in Chapter 3.

Basic Systems Flow—COM Systems

The COM system involves five steps:

- COM conversion.
- Processing of the exposed microfilm.
- Formatting the microimages.

- Creation of additional copies of the microimages for distribution and security purposes.
- Referencing the microimages.

COM Conversion

The conversion of data from a digitized, computer-processible format to ordinary alphabetic and numeric characters on microfilm may occur as either an on-line or as an off-line operation. In the *on-line mode*, data is transmitted directly from the computer's central processing unit (CPU) to the COM recorder. In the *off-line mode*, data transmission occurs from the input/output (I/O) device (i.e., the magnetic tape drive, or the magnetic disk drive) to the COM recorder.

Through internal programming, the magnetic data is converted to a series of electrical impulses and finally to a light source that either "writes" the alphabetic and numeric characters directly on the microfilm, or projects them on the face of a Cathode Ray Tube (a CRT) where they are photographed by a special-purpose microfilm camera that is equipped with plug-board logic.

COM output is initially either 16 or 105 mm wide roll microfilm, typically is continuous 1,000-foot lengths.

Processing

Depending upon the COM recorder used, the processing of the exposed microfilm will either be a separate operation using the wet chemical process earlier described for Source Document Micrographics, or else be a function of the COM System itself, using a dry process.

Inspection and Formatting

After processing, the COM-generated microfilm is inspected to assure satisfactory quality, and any subsequent formatting is performed (e.g., the 105-mm-wide roll film is cut to form individual 148 × 105 mm microfiche). Typically, 16-mm roll film will not be reformatted until after any duplication has occurred.

Duplication and Distribution

The creation of duplicate roll microfilm or microfiche of COM-generated microimages is accomplished in the same manner described earlier for Source Document Micrographics, using either roll or microfiche duplicating equipment, and producing duplicates on either silver halide microfilm or diazo or vesicular films. Following the creation of duplicate copies for distribution or security purposes, the 16-mm-wide microfilm is cut to 100- or 150-foot lengths (as applicable) and either rewound on open reels or inserted into cartridges or cassettes, if such formats are to be used.

Reference

Enlargement of the COM-generated microimages for direct viewing or reproduction purposes is accomplished through the use of special-purpose readers and reader-printers, which operate in the manner described earlier for Source Document Micrographics.

COM technology and the various aspects of the COM Systems will be discussed in greater detail in Chapter 8.

2

Microforms

The selection of the proper storage media and index is the key to the design of a cost-efficient, responsive, user-oriented information storage and retrieval system. The way in which the individual records are organized, and the order in which they are maintained in the file, will determine how a manual or a mechanized retrieval system may manipulate the storage media—the microfilm, magnetic tape, paper records, and the like—to access a particular item of information. Unless the information is organized and filed in a manner that will facilitate its storage and retrieval, and indexed so that any given record may be quickly and accurately located, the system will invariably fail.

Early applications of microfilm were generally unsatisfactory to many users because the alternatives for both storage and indexing were severely limited. Thus, many users were reluctant to utilize microfilm for any but their least referenced records that could be stored on continuous roll film and be indexed in a sequential alphabetic, numeric, or chronological order. Today, this situation has changed. The desire to use microfilm for both *active* and *inactive* information storage and retrieval has led to the development of a number of alternative microfilm formats, each providing an efficient, economical solution to a specific type of filing and retrieval requirement. These formats may be classified in two categories:

• **Unitized.**   A microfilm format that is planned and prepared as one complete unit or sub-division of information, without reference to any unrelated or extraneous materials. The unitized microfilm format may be compared to a file folder that groups together all data relating to a particular subject, event, person, or transaction, and that contains no information or records that do not relate to that person, and so on. Examples of unitized microfilm formats (or, as they are properly called, *microforms*), are microfiche and microfilm aperture cards and microfilm jackets.

• **Non-unitized.** Microfilm that contains unrelated records and units of information. Frequently roll and cartridge microfilm will be nonunitized since they will contain a variety of unrelated information on the same continuous length of microfilm.

It is more costly and time-consuming to prepare microfilm records in a unitized format than in a nonunitized. Therefore, before specifying a unitized format, one should assure that the system's storage and retrieval requirements justify the additional labor and expense.

As a general rule, unitized formats should be considered whenever groups of records are referenced as a unit, as in the case of personnel files or financial statements, or when frequent updating is required, as in the case of active research files, credit files, and patient medical folders.

Nonunitized microforms are best suited for applications in which little or no updating is anticipated.

NONUNITIZED MICROFORMS

16- and 35-mm Roll Film

The greater majority of source document microfilm applications involve the use of 16- and 35-mm silver halide negative microfilm, in continuous 100- and 200-foot lengths mounted on open reels, as the recording media (Figure 2-1). After processing the roll microfilm may be segmented to form various microforms, such as aperture cards or microfilm jackets, or it may be retained in its original roll format.

Image Orientation

Microimages may be arranged on 16- and 35-mm roll microfilm in three methods (called "orientations"):

- Simplex.
- Duplex.
- Duo.

In the *Simplex Orientation,* which is the most common method, the individual microimages are sequentially arranged in a single row on the microfilm. Indi-

Figure 2-1   16-mm roll film. Courtesy 3M Company.

vidual simplex-oriented microimages, in turn, may be filmed in one of two *modes:*

- **Comic Mode** (Figure 2-2). The microimages are arranged side-by-side along the length of the microfilm.
- **Cine Mode** (Figure 2-3). The microimages are arranged one under the other in a head-to-foot sequence across the width of the microfilm.

In the *Duplex Orientation* (Figure 2-4), the front and back sides of the source documents are simultaneously photographed and appear side by side, in a Cine Mode only, across the width of the 16- or 35-mm microfilm. This format is typically used for the microfilming of cancelled checks and other less than letter-size two-sided documents.

In the *Duo Orientation,* the individual microimages appear in two rows, in either a Comic Mode or a Cine Mode. The microfilm is run through the camera twice. During the first pass, the microimages are sequentially positioned on one-half of the film's width. When the end of the microfilm is reached, it is removed from the camera, turned around, and reinserted into the microfilm camera. The microfilming then resumes, with the microimages appearing in sequential order on the other half of the film's width. In effect then, the duo format provides two rows of microimages running up one side of the continuous

Figure 2-2   Simplex orientation—comic mode.

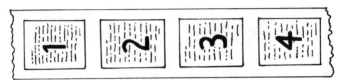

Figure 2-3   Simplex orientation—cine mode.

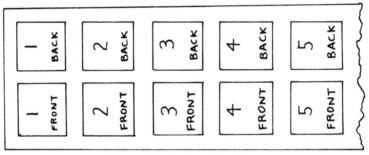

Figure 2-4   Duplex orientation.

microfilm and back down the other. Figures 2-5 and 2-6 illustrate the Duo Orientation with microimages arranged in a Comic and a Cine Mode respectively.

Advantages

As a microform, 16 and 35 mm roll microfilm has the following advantages:

- Maximum storage capacity—roll microfilm provides a 98% reduction in space and equipment requirements over hard copy.
- Least costly microform to prepare.
- Ideally suited for the rapid, inexpensive reproduction, in either hard copy or microfilm, of the entire 100- or 200-foot roll, or of major segments thereof.

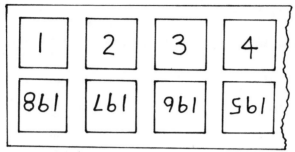

Figure 2-5   Duo orientation—comic mode.

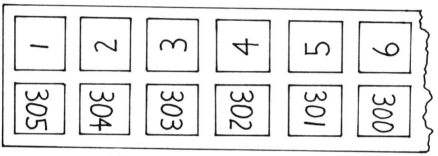

Figure 2-6   Duo orientation—cine mode.

- Affords a secure and inviolate storage media. The problems of lost or misfiled records are minimal when they are microfilmed and retained in a continuous roll format.

Disadvantages and Limitations

Conversely, however, roll microfilm has the following offsetting disadvantages and limitations:

- It is expensive, time-consuming, and difficult to update. Updating requires either (a) the splicing of the additional microimages into the roll in their proper location, or (b) the creation of a trailer roll.
- Generally, microimages on continuous roll microfilm are slow and costly to reference.
- Many persons experience difficulty threading roll microfilm through readers and reader-printers. Consequently, damage to the microfilm often results.

Cartridges and Cassettes

To speed up and simplify retrieval, as well as reduce the incidence of damage to the roll film as the result of rough handling or improper threading into readers and reader-printers, roll microfilm may be housed in cartridges and cassettes.

• The *cartridge* (Figure 2-7), as defined by the National Micrographics Association "is a container, designed to be inserted into microfilm readers, reader-printers, and retrieval devices." The end of the continuous 100- or 200-foot roll microfilm is adhered to a single spool located in the interior center of the cartridge. The microfilm is then wound around this spool, and a magnetized "tip" is added to the unsecured end of the microfilm. When the cartridge is inserted into a suitable reader or reader-printer, the magnetized tip is automatically drawn through the various film gates, self-threading the microfilm rapidly, accurately, and in a damage-free manner. After reference the microfilm automatically rewinds into the cartridge and may be readily removed from the reader or reader-printer without anyone touching or handling the microfilm in any manner.

• The *cassette* (Figure 2-8), has two cores: a feed and a take-up. The trailing end of the continuous microfilm roll is adhered to the feed core. The microfilm is wound around this core, threaded through the cassette, and the leading end is adhered to the take-up core. When reference is required, the microfilm advances from the feed to the take-up core automatically. Since the cassette is a self-contained unit, there is no need to rewind the microfilm after reference. All one need do is lift the cassette out of the reader or reader-printer. Although cartridges

Figure 2-7   Cartridges. Courtesy 3M Company.

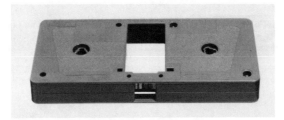

Figure 2-8   Microfilm cassette.

and cassettes do simplify handling, and reduce the possibility of damage to the microfilm, the disadvantages noted earlier for continuous roll microfilm are still present. Cartridge- and cassette-contained continuous roll microfilm is difficult to update, and the retrieval of individual microimages on the 100- or 200-foot continuous roll is still a relatively slow proposition.

UNITIZED MICROFORMS

There are five primary unitized microforms:

- Micro-opaques.
- Microfilm jackets.
- Microfiche.
- Microfilm aperture cards.
- Ultrafiche.

Micro-Opaques

Micro-opaques are available in two basic formats: cards and rolls. Each of these are prepared as a continuous contact printing process.

Microcard

Microcards (see Figure 2-9) are produced at reduction ratios of between 17× and 23× on cards measuring 5 × 3, 6 × 4, or 8 × 5 inches, utilizing one of the following methods:

- By means of a contact print of a negative microfilm roll on photographic paper.
- By the use of a step-and-repeat camera that produces a negative microfiche.

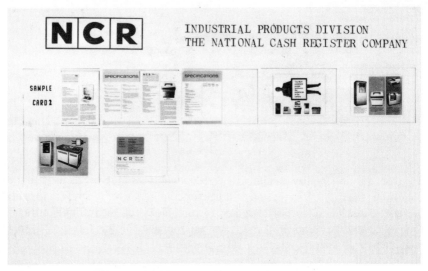

Figure 2-9   Microcard. Courtesy NCR Company.

The negative microfiche is then used to produce the microcard, via the contact printing process.

Microcards, which may be printed on either a single or both sides, are used as a publishing medium in those applications in which only references and note taking is involved. For example, copies of United States Patents have for many years been available in microcard format, as have molecular structures.

Since microcards are opaque, they must be viewed by reflected rather than transmitted light. Therefore, a special purpose microfilm reader is required for use with micro-opaques.

Until recently, it was not possible to reproduce enlarged copies of micro-opaque images in a hard copy format. There are now a limited number of reader-printers that accept micro-opaques, however.

Micro-Print Cards

The ink-printed *micro-print card* is similar to the microcard in both appearance and format. Micro-print cards are produced as follows: An offset printing plate master that contains microimages from 16- or 35-mm roll film is prepared and mounted on an offset printing press. Then it is merely another printing process, similar to that used in conventional book publishing. The low-cost microfilm-to-offset process permits the production of inexpensive, large volume, reports, books, periodicals, and other records required primarily for reference purposes.

Both the microcard and the micro-print card is best suited for use with completed and closed files—those that require no further revision or updating. If such revision and updating is required, and a micro-opaque is desired, a micro-opaque in roll form should be considered.

Micro-Opaques in Roll Form

This microform too is prepared as a continuous printing operation from roll microfilm. Negative continuous roll microfilm is used to prepare a positive microfilm image on 16- or 35-mm wide photographic paper tape with a pressure-sensitive adhesive backing. After processing, the positive microform is simply cut to the proper length and pressure-applied to ordinary index cards.

Micro-opaques in roll form possess excellent add-on capabilities, since any particular micro-opaque card may be updated without refilming the entire file. However as in the case of microcards and micro-print cards, only a limited number of reader-printers are available to produce enlarged, hard copy reproductions of individual microimages.

UNITIZED NEGATIVE MICROFORMS

Microfilm Jackets

The microfilm jacket (Figure 2-10) provides an efficient, economical means of storing and retrieving records that must be periodically updated and duplicated. Microfilm jackets are composed of two clear sheets of cellulose acetate separated into 16- and/or 35-mm wide channels that are used for the storage of short lengths of microfilm containing related images.

Available in a wide variety of sizes—5 × 3, 6 × 4, 8 × 5 inches, and tab card size—microfilm jackets may be updated to include new and revised records merely by inserting additional lengths of microfilm containing such microimages into the next available channel, and by removing and discarding any frames of microfilm that contain obsolete microimages.

The microfilm jacket's transparency permits the reproduction of enlarged copies of any microimage filed in the microfilm jacket without first having to remove the microfilm from its microfilm jacket. It also permits the duplication of the entire microfilm jacket, using a suitable fiche-to-fiche duplicator.

Microfilm jackets are ideally suited for those applications that require the following:

- A unitized, negative microform.
- The ability to add additional microimages to the microform after its initial

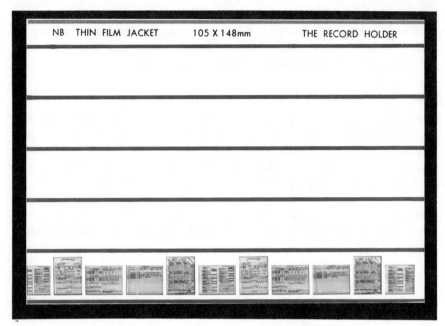

Figure 2-10   Microfilm jacket. Courtesy N.B. Jacket Corporation.

preparation, or to supersede and/or delete individual or groups of microimages.

- Capability of rapid, on-demand duplication of selected, individual microimages or of the entire microfilm jacket.
- Moderately low-distribution requirements.

Advantages of Microfilm Jacket Format

Microfilm jackets have the following advantages:

- They can be updated to add new microimages and to delete existing ones.
- They are readily reproducible for reference and distribution purposes.
- They afford the microimages protection from dirt and scratches.

Disadvantages of Microfilm Jacket Format

The main disadvantage of the microfilm jacket format is its slowness of preparation, especially if only one or two microimages are to be inserted into a single microfilm jacket.

Microfiche

Microfiche, or "Fiche" as it is commonly known, is a sheet of film, generally 6 × 4 inches in size, that contains multiple microimages arranged in a grid pattern in the same way that the days of the month appear on a calendar (see Figure 2-11).

Production of Microfiche

Microfiche may be viewed as a printing process for publishing in microimage form, where a moderate-to-large number of distribution copies are required. The production cycle involves two distinct steps:

- The preparation of a microfiche master.
- The preparation of duplicate microfiche from that master for distribution purposes.

The microfiche master may be created in one of the four following ways:

• **Stripped-Up from Continuous Roll Film.**   Planetary or rotary camera-produced roll microfilm is cut to approximate 6-inch lengths, coated with an adhesive backing and stripped onto a transparent sheet of film.

• **Duplicated from Microfilm Jackets.**   A microfilm jacket is duplicated using a suitable fiche-to-fiche duplicator to generate a microfiche master.

• **Photographically from Source Documents using a Step-and-Repeat Microfilm Camera.**   The step-and-repeat microfilm camera is a precision

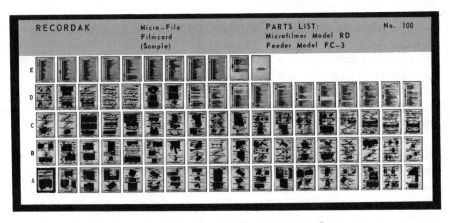

Figure 2-11   Microfiche. Courtesy Eastman Kodak Corporation.

planetary camera that photographs each source document and places its microimage in exactly its proper and precise grid position on a 105-mm-wide microfilm. Step-and-repeat microfilm cameras operate in a fashion that is similar to a typewriter. As each microimage is positioned on the film, the camera "steps" to the right to the next recording position. When an entire row of microimages has been positioned, the camera automatically advances to the first, left-most filming position in the next row and "repeats" the procedure.

• **Produced from Computerized Data via the COM Process.** This technique, which will be examined in Chapter 8, produces a microfiche master as the result of digital data being converted to ordinary alphabetic and numerical characters by a COM recorder and subsequently being transferred as microimages onto a 105-mm-wide film.

Duplicate microfiche for distribution and security purposes are then prepared from the Microfiche Master, using a suitable fiche-to-fiche or roll-to-fiche duplicator.

Selection of Microfiche or Microfilm Jackets

According to Richard W. Batchelder, a Fellow of the National Micrographics Association and one of the world's leading micrographics technicians, "from an economic standpoint, the break-even point of microfiche as opposed to microfilm jackets is reached whenever the total copies required for distribution and security purposes exceeds 15 copies. Anything less than that amount," Batchelder explained, "and you will be economically ahead if you use the microfilm jacket format."

Standardization of Microfiche

Some 20 years ago, when microfiche began to become widely used in the United States, there were no standards with respect to the format of this microform. Each manufacturer and user was left quite on his own to design the microfiche that most closely approximated his needs. Consequently, one found that reduction ratios ranged from 18 to 32×. Documents were arranged from lower left to top right in some instances and from top left to lower right in others. Indexing was nonuniform, as was the placement of microimages on the microfiche itself.

In an attempt to correct these discrepancies, and to provide for the interchange of both microfilm hardware as well as the microfiche themselves, two major standards—the American National Standards Institute's (called the ANSI Standard for short), and the Council on Scientific and Technical Information's (better known as the COSATI Standard) were developed. Generally, the ANSI Standard found greatest acceptance among organizations in the private sector of the economy, which were not required to submit microfiche records to the government in satisfaction of a government contract. The COSATI, which was mandatory for all federal agencies was fairly well restricted to such governmental bodies and their contractors and subcontractors. Subsequently, the ANSI and the COSATI Standards were combined into the NMA Microfiche Format which is applicable to all microfiche created by other than the COM Processes—that is, to all microfiche containing images created initially from hard copy-based source documents.

The NMA Standard provides for the following:

- Size of the microfiche: 148 × 105 mm (nominally 6 × 4 inches)
- Standard reduction ratio: 24×
- Image grid: 98 frames total, arranged in 7 horizontal rows, each containing 14 frames. In addition, a title area is provided at the top of the microfiche for a descriptive index, identifying the individual microfiche's contents.
- Grid index and page sequence: Each row is alphabetically identified from top to bottom beginning with "A." Columns are identified numerically from left to right beginning with "1." Pages are to be arranged in a left to right sequence.

Unlike Source Document-generated microfiche, there is no single standard for COM-generated microfiche. Two reduction ratios are provided for 42× and 48×, and the image grid is adjusted accordingly. The grid index and the page sequence for COM-generated microfiche is the same as that for Source Document-generated.

Updateable Microfiche

It should be noted that the newest concept in microfiche is the *updateable microfiche*. This type of microfiche enables the original microfiche master to be updated at any time to alter or add to the microimages contained therein.

Several micrographics manufacturers are in various stages of development on this new technique, but only one, A. B. Dick/Scott's System 200, is currently being marketed. The updatable microfiche concept involves the exposure and processing of one or more individual "grid positions" on the Microfiche Master, rather than the entire microfiche. It is therefore possible to update the microfiche by adding new records as they are created and received.

A Health and Beauty Aids Manufacturer has successfully installed a System 200 in their Credit and Collections Department. At the installation of the new system, all Customer Credit History Files were converted from hard copy to updatable microfiche format. Now, as new Mercantile Credit Reports, internal analyses to determine the customer's creditworthiness and the like, are received, they are added to the updatable microfiche in a simple microfilming operation. Should any record included on the updatable microfiche be superseded, a large red dot is placed directly in the center of the superseded microimage using a magic marker. An individual referencing that microimage will notice the superimposed red dot and will then look at the following images for the superseding microimage.

Advantages

- Microfiche provides a quick and economical means of preparing, duplicating, and mailing multipaged reports and files.
- Microfiche will reduce dramatically the file space and equipment required for the maintenance of files, books, reports, and similar records.
- Unlike the original hard copy, microfiche will not tear and if properly processed and stored, will not deteriorate or yellow with age.

Limitations

- With the exception of the single updatable microfiche system, microfiche prepared by step-and-repeat process cannot be updated.
- If the size of the files being converted to microfiche does not approximate the microfiche's capacity, film will be wasted.
- Experience indicates that a greater incidence of loss and misfiling will occur if microfiche is used than with one of the nonunitized formats.

Microfilm Aperture Cards

The microfilm aperture card (Figure 2-12), is a standard 80 column tab card that measures 7⅜ × 3¼ inches. Columns 1 through 52 are available for key punching relevant information as a means of identifying and retrieving the card. Columns 53 through 80 contain a die-cut aperture in which the chip of microfilm containing one ore more microimages is placed.

Microfilm aperture cards are well-suited for applications in which engineering drawings and supporting data, such as bills of materials, drawing lists, and engineering change notices, must be filed, duplicated, distributed, and updated.

Production Method

Original source documents are microfilmed using a planetary microfilm camera that has been equipped with a 2-inch pull-down feature (an attachment that advances the microfilm 2 inches after each exposure). After processing, the microimages are cut to individual chips, an adhesive coating is applied to all four edges, and the microfilm chip is adhered to the tab card over the die-cut aperture.

Should duplicate microfilm aperture cards be required for distribution or security purposes, such copies may be quickly and inexpensively prepared by a card-to-card duplicator (described in Chapter 3). The original microfilm aperture card is placed directly atop another microfilm aperture card containing a pre-mounted, blank diazo film chip over the aperture. As the matched cards pass through the card-to-card duplicator, the original microimage is reproduced on the duplicate aperture card's blank film chip.

Image Orientation

Engineering drawings and other records that are larger than legal size (8.5 × 14 inches) are generally arranged on the microfilm aperture card in a Comic Mode.

When the microfilm aperture card is used to store microimages of records that are up to legal size, it is possible to include up to eight microimages in a single microfilm aperture card (Figure 2-13). Such multiple microimages may be arranged in either a Comic or a Cine Mode.

A modification of the microfilm aperture card provides for the inclusion of channels across the aperture area, thus combining the storage features and capabilities of the microfilm jacket, with the convenient size and key punching and mechanical retrieval features of the microfilm aperture cards (Figure 2-14). If such a format is used, the channels may be positioned so that the microimages are arranged in either a Comic or a Cine Mode.

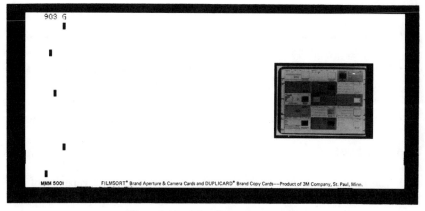

Figure 2-12   Microfilm aperture card.

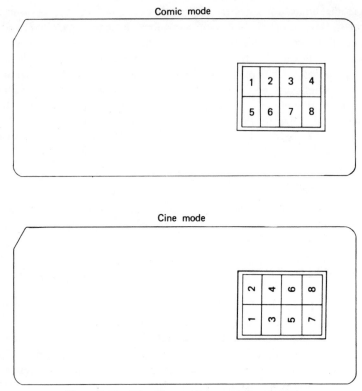

Figure 2-13   Microfilm aperture card, 8-up format.

Advantages

- The creation of duplicate microfilm aperture cards is a rapid, simple, and relatively inexpensive procedure.
- Since either the original microfilm aperture cards or a "slave deck" of such cards may be key punched to include key data concerning the microimages contained, high-speed machine sorting and retrieval is possible.
- Other microforms may be created from the microfilm aperture card.

Disadvantages

- The microfilm aperture card provides the lowest storage capacity of any microform. However, it does provide a far greater density than does hard copy.

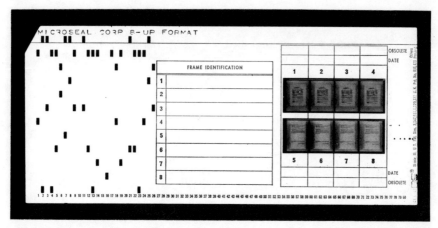

Figure 2-14   Microfilm aperture card with channels. Courtesy Microseal Corporation.

• There is a high incidence of loss or misfiling compared to nonunitized microforms.

Ultrafiche

Ultrafiche is a term that is applied to microfiche that has been produced at reduction ratios of 90× or more. Such a format affords extremely high storage density; a single 6 × 4 inch ultrafiche may contain nearly 10,000 microimages.

Ultrafiche is primarily used for the publication and distribution of equipment repair manuals, parts and service manuals, and other micropublishing applications, since its extremely high storage density allows many thousands of microimages to be reproduced on a single fiche, minimizing the inconvenience and the possibility of loss that would occur were multiple microfiche prepared at standard reduction ratios.

The ultrafiche (see Figure 2-15) used with a suitable reader will produce a clear legible viewing image.

Advantages

• The ultrafiche has the maximum storage capability of all microforms.
• The preparation of duplicate ultrafiche for distribution and security purposes is a rapid, inexpensive procedure.

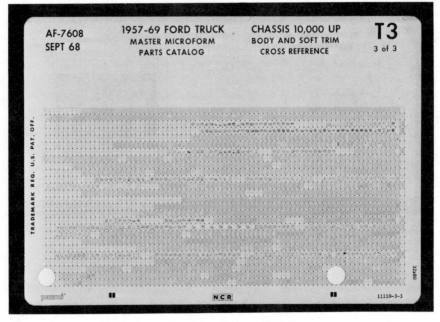

Figure 2-15   Ultrafiche. Courtesy NCR Company.

Disadvantages

- The master ultrafiche is extremely expensive to create. Depending upon the number of documents and the system involved, a single ultrafiche master will cost between $700 and $5,000 to prepare.
- Updating is a costly, slow process.
- It is difficult to subdivide an ultrafiche to provide for the reference and use requirements of those users who may not require all the information contained therein.

3

Micrographics Hardware

Micrographics hardware consists of equipment that has been developed for the express purpose of creating, duplicating, retrieving and referencing unitized and nonunitized microforms. Such hardware may be classified into six distinct categories:

- Source Document Cameras and COM Recorders
- Processors
- Duplicators
- Readers
- Reader-Printers
- Automated Microfilm Information Storage and Retrieval Systems

SOURCE DOCUMENT CAMERAS AND COM RECORDERS

As described in Chapter 1, there are three types of Source Document Cameras, each designed for the microfilming of hard copy source documents of specific formats and dimensions.

Rotary Microfilm Camera

The *rotary microfilm camera* (Figure 3-1) is designed for the high-speed microfilming of hard copy source documents that are generally legal size (8.5 × 14 inches) or smaller and are individual sheets of paper.

Figure 3-1   Rotary microfilm camera. Courtesy 3M Company.

The rotary microfilm camera is extremely simple to operate, so simple in fact, that one manufacturer once trained a Gibbon to operate one model during a trade show. Simply load the microfilm into the camera, turn on the power, and begin feeding documents into the camera either manually or mechanically. The exposure is preset, so no special operator skills or extensive training is required.

Figure 3-2 is a schematic representation designed to illustrate the operation of the typical rotary microfilm camera. Although it does not depict the actual operation of any one specific camera, it is representative of all rotary microfilm cameras.

When the camera operator turn the power on, the rotary microfilm camera's internal flood lights (A) ignite. The hard copy source documents are then fed into the rotary microfilm camera, either manually or by a mechanical feeding attachment, and enter at point (B). Here, the document is picked up and pressed against a continuously moving belt that transports it to the interior of the camera. When the forward (or leading) edge of the document passes in front of the forward sensor (C), a signal is transmitted to the film take-up unit (D) and the microfilm begins to move across the camera's lens synchronously with the speed of the hard copy document passing below. Thus, the moving document is filmed by moving microfilm, thereby reducing the likelihood of a distorted microimage, and enabling high-speed microfilming to occur. The microfilm will continue to move and photograph the area directly beneath the camera's lens, so long as the rear (or trailing) edge of the document does not pass by the sensor that is located at Point (E). When this does happen, a signal is transmitted to the Film Take-Up Unit and the film ceases movement until it is again activated by the next document's

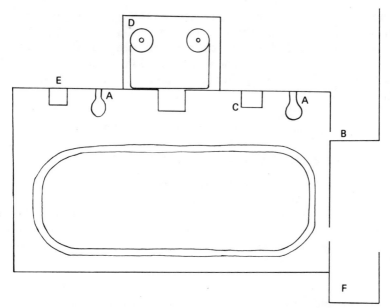

Figure 3-2 Schematic representation of rotary microfilm camera.

leading edge passing in front of and activating the sensor at point C. Meanwhile, the continuously moving belt has transported the document through the camera and deposited it into the Output Hopper (F) awaiting its removal.

The output of the rotary microfilm camera may be compared to that of the fixed focus box camera. It is generally legible, and will prove satisfactory in the majority of applications. However, since its resolution (or legibility) can not be improved by such factors as varying the focal length, a rotary microfilm camera-generated microimage may not prove satisfactory when it is necessary to assure, as in the case of engineering drawings, a high degree of image quality in both the viewing and reproduced images. In such instances, the planetary camera will generally prove the more feasible choice.

The rotary microfilm camera, available in either a portable or as a stationary installation, is the workhorse of office microfilming operations. With actual operating speeds ranging from 600 hand-fed letter-size documents to as many as 4,500 mechanically fed checks in a single hour, and almost trouble-free operation the rotary microfilm camera is well suited for the filming of correspondence, forms, checks, and other unbound source documents.

A major advantage of the rotary microfilm camera is its ability to accept a document of any length, provided its width does not exceed a specified limit, which is usually 12 inches. As noted earlier, so long as the trailing edge of the document does not pass in front of the camera's rear sensor, the filming process will continue regardless of how long the document is. This enables the rotary microfilm camera to be used for such records as oversized ledger sheets, burst computer print-out (unburst print-out will exceed the width limitations and so could not be microfilmed by a rotary microfilm camera, and various graphs, charts, and worksheets.

The major limitations of the rotary microfilm camera are its fixed focal length, its width limitations, and its inability to accept bound or thick records.

Rotary Cameras for Unburst Computer Output

A rotary microfilm camera specifically designed for the continuous, high-speed microfilming of unburst computer output provides an alternate means of microfilming such records (Figure 3-3). Operating in a simplex made at a 24X reduction ratio, these special purpose rotary microfilm cameras will film computer print-out or any other unburst continuous forms at speeds of approximately 150

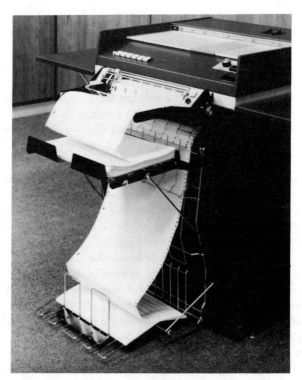

Figure 3-3   Rotary microfilm camera—unburst computer output. Courtesy 3M Company.

linear feet (approximately 160 pages) per minute. For applications where hard copy is requested for only the first portion of the record's retention, its conversion to microfilm by such a special purpose camera may well prove the most efficient, economical alternative.

Planetary Camera

The Planetary Camera (Figure 3-4) is designed for the microfilming of bound books and documents, records as engineering drawings that require high resolution and image quality, records that are too large to put into the rotary microfilm camera, and those records so valuable that one dares not risk its possible damage while being transported into and through the microfilm camera. Figure 3-5 is a schematic representation illustrative of the operation of planetary cameras.

The records to be microfilmed are laid on the flat-bed surface of the planetary camera (A) positioning it so that all four corners fall squarely within the appropriate angle marks. This assures that the record is centered underneath the camera lens and is squarely aligned. The focal length adjustment wheel (B) is then turned to either lower or raise the camera unit (C) on its shaft; thus varying the reduction ratio at which the document will be microfilmed (Figure 3-6). The power is then turned on, the spotlights (D) illuminating the document on the flat-bed. The camera operator next checks the exposure meter (E) on the front panel, adjusting

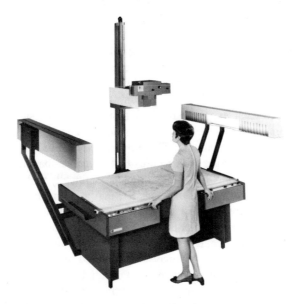

Figure 3-4   Planetary camera. Courtesy ITEK Corporation.

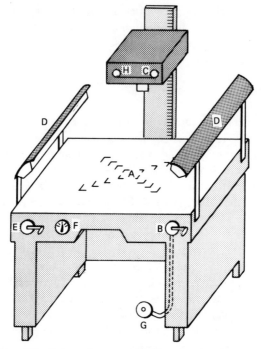

Figure 3-5   Schematic representation of a planetary camera.

the lights by turning the light adjustment wheel (F) until the proper setting is obtained. The camera operator then steps on the foot switch (G), which signals the camera unit to take a single exposure. After the exposure, the film take-up unit (H) advances to the next filming position.

Naturally, since each exposure requires that the operator step on the foot pedal, the operating speed of the planetary camera is considerably slower than that of the rotary microfilm camera. The writer's experience is that an experienced camera operator working with records that have been batched by size and reduction ratio (thus reducing the time required to continually realign records and move the camera unit up and down its shaft), may reasonably be expected to average 500 documents per hour. If, however, the records to be filmed are of varying size and are to be filmed at different reduction ratios, the output can be expected to decrease significantly.

If the rotary microfilm camera may be compared to the box camera, the planetary camera may be compared to the Portrait Camera. It is a precision camera that films a stationary object from a stationary position, thereby eliminating any problem of streaking or other conditions resulting from the film and the

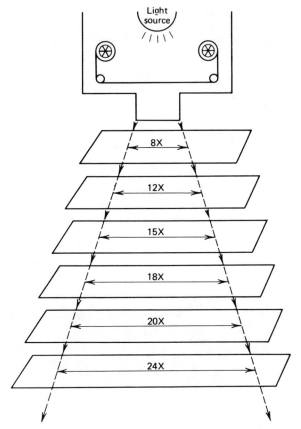

Figure 3-6 How focal length determines reduction ratio.

document not moving synchronously. Its focal length can be adjusted to vary the reduction ratio of the microimage. The lighting intensity and the line voltages can be modified to reduce the incidences of over- or under-exposure.

Where microimages of engineering drawings and other records are to be cut to size and mounted in microfilm aperture cards, a special attachment called a "Two-Inch Pull Down Bar" may be built into the planetary camera. This attachment automatically increases the distance that the microfilm advances between each image, so that a 2 inch clear area is provided, thus assuring adequate space for cutting the microfilm into individual microimages or "chips," and for gluing and mounting each chip over the aperture that has been die-cut in the tab card.

The major advantage of the planetary camera is its adjustable focal length. This feature provides for varying reduction ratios; the ability to compensate for excessively light or dark background or data, by varying the intensity of the lighting on the document; finally, its ability to film bound books and records, as well as oversize records, drawings, and other documents without having to alter their original format.

The planetary camera's primary disadvantages, compared to the rotary microfilm camera, are its slower operating speed, its generally higher cost, and its complexity, which requires that it be operated by a trained person.

Step-and-Repeat Camera

The step-and-repeat camera (Figure 3-7) is a special-purpose planetary camera that is used solely for the production of microfiche. The individual records are filmed in sequence using the step-and-repeat camera. In turn, they are positioned in a left-to-right, top-to-bottom sequence on either a 105-mm-wide continuous roll microfilm or on a 148 × 105 mm sheet of negative microfilm.

Figure 3-7   Step-and-repeat camera. Courtesy 3M Company.

The actual filming operation of the step-and-repeat camera is similar to that of the planetary camera. The camera operator again positions the source document to be filmed between the appropriate angle marks on the camera's flat bed. After adjusting the focal length to provide the desired reduction ratio, the power is turned on. Checking the exposure meter and making any required variations in the intensity of the light flooding the document, the camera operator then steps on the foot pedal. This activates the camera, which takes a single exposure and immediately advances the film within the Camera Unit itself to the next filming position.

As described in Chapter 1, during our discussion of the Source Document Micrographics process, the filming sequence of the step-and-repeat camera resembles that of a typewriter. After each exposure, the camera "steps" to the right, stopping at the next filming position. When the prescribed number of exposures have been positioned in one horizontal row, the camera returns to the left hand most position of the next lower horizontal row, and the stepping-to-the-right filming continues.

COM Recorder

The COM recorder provides for the preparation of microimages of data that are stored in a computer-processible format (magnetic coding, punched holes, etc.), without first having to print such data out in a hard copy format. In Chapter 8, the operation of the COM recorder will be examined in detail. At this point, we need only mention that, depending upon the system involved, data may be transferred either directly from the computer's central processing unit (CPU) or from a tape or disk drive to the COM recorder, where it is either displayed on the face of a Cathode Ray Tube (CRT) and then microfilmed, or is transferred ("written") directly on the microfilm itself.

Evaluation Criteria

In determining which type of microfilm camera(s) will be best suited to a specific organization's needs, the following data must be determined:

- The format and dimensions of the documents to be microfilmed.
- Filming modes required.
- Degree of resolution required.
- Reduction ratio to be utilized.
- The need for portability.
- Indexing requirements.
- Form of output desired.

These varying requirements should be itemized on an ordinary columnar pad. When all the evaluative criteria and requirements have been determined and listed, it becomes a relatively simple analysis to determine the features that the cameras must have. Assume, for example, that it is determined that the following features and capabilities are required:

- Ability to film single pages up to 8.5 × 11 inches
- Simplex mode.
- The image must be clear and readable. A high degree of resolution, however, is not essential.
- The reduction ration must be either 20 or 24X.
- There is no need for a built-in indexing system, such as Code Line Indexing.
- Roll microfilm in continuous 100-foot lengths is the desired output.

These requirements may then be compared to the specifications of the various model microfilm cameras available on the market to determine which will satisfy the organization's projected needs. A further evaluation, and actual testing of those cameras that appear to satisfy such needs, will lead to the selection of the camera that best satisfies the organization's projected reference and use requirements.

MICROFILM PROCESSORS

The microfilm processor (Figure 3-8) develops the latent image once it has been created on the silver halide microfilm, "fixes" it by making it impervious to future exposure to light, and washes the chemicals off the surfaces of the processed microfilm.

Microfilm processors that are marketed today range from small, relatively inexpensive units having slow output speeds that are designed for organizations whose total daily microfilm processing requirements average upward of 1,500 linear feet, to large, costly, high-speed units designed for use by commercial microfilm processing laboratories and organizations that process tens of thousands of linear feet of microfilm every day. Regardless of their size, output speed, and cost, however, all microfilm processors operate under normal room lighting, thereby eliminating the need for a photographic darkroom and the various equipment and controls that are associated with conventional photographic processing.

Figure 3-9 is a schematic representation of a typical microfilm processor. The exposed microfilm is attached to a long blank film leader and is wound onto the

Figure 3-8   Microfilm processor. Courtesy 3M Company.

microfilm processor's supply reel. After the film has been wound onto the supply reel, another long blank film leader is attached to the leading edge of the microfilm. The leader is then threaded through the processor's various vats and through the dryer. It is then inserted into the take-up reel, and the power is turned on. The microfilm then passes in a steady, continuous path through the developer vats, where the latent microfilm image is produced. It next goes through a series of water baths to remove all traces of developer from the surfaces of the microfilm, and then onto the vats that contain *hypo* (sodium thiosulfate penthydrate), a solution that "fixes" the microimage and causes it to be unaffected by subsequent exposure to light. Then, after passing through another series of water baths to remove the hypo, the microfilm is transported through the dryer and is wound onto the Take-Up Reel.

As a general rule, the in-house processing of microfilm cannot be cost-justified unless an organization processes at least 250,000 linear feet of micro-

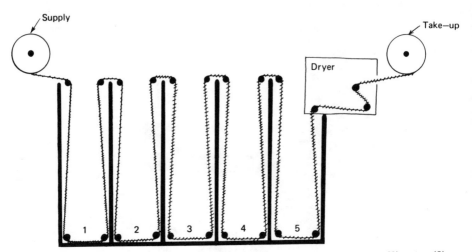

Figure 3-9   Schematic representation of a microfilm processor: (1) developer, (2) water, (3) water, (4) hypo, (5) water.

film annually. Below that level, the savings that result from purchasing microfilm *without* processing included, will not achieve a three year payback of the microfilm processor's cost, taking the average processing costs per 100 linear feet of microfilm processed into account. There are, however, a number of operational requirements that will normally override the economic factors and necessitate an organization acquiring its own in-house microfilm processing capability:

- The nearest commercial microfilm processing laboratory may be located at such a distance from the organization that mail may prove to be the only feasible means of transporting the microfilm to and from the commercial processing laboratory. Rather than wait several days between the time the exposed microfilm is mailed and its processing and return, and in order to minimize the possibility of loss or damage, the organization may choose to install and operate its own microfilm processor.

- Records of a proprietory, highly confidential nature may be scheduled for periodic microfilming. Rather than risk unauthorized access to, or the loss of, such information, the decision may be made to process microfilm internally. Although experience has proven that the possibility of loss or compromise is virtually nil if one deals with a reputable commercial microfilm processing laboratory, many organizations feel much more secure when such proprietary, highly confidential information, in microfilm form, is processed within its own facilities, by its own personnel, on the organization's own equipment.

- The microfilm may be of records that will pass out of the organization's custody shortly after their microfilming, and the time is just not available to use a commercial microfilm processing laboratory. Examples of such records would be cancelled checks that are presented for payment at one bank and must be microfilmed before being forwarded to the clearing house for their eventual return to the banks upon which they were drawn. Another example would be the various stocks and bonds that are microfilmed for security and reconstruction purposes every time they are either deposited into or withdrawn from the safekeeping vault. Microfilm copies of such records must be processed as quickly as possible, so that they may be inspected before the original source documents leave the organization's custody, so that if any refilming is necessary, the original source documents will still be available.

Evaluation Criteria

In selecting a microfilm processor for in-house operation, the following data must first be determined:

- The width of the microfilm that will be processed (16, 35, 70, 105 mm).
- The length of such microfilm (100, 150 linear feet, etc.).
- The volume of microfilm to be processed daily.

This will define the operating capabilities that must be satisfied by the microfilm processor. By comparing these requirements to the specifications of the various models, it will be possible to identify those microfilm processors that are capable of satisfying the organization's identifiable needs. It then becomes a straightforward matter to review each of these processors, taking into account such variables as their size, plumbing, venting, and electrical requirements, ease of loading and removing chemicals, and the like, to determine which will prove to be the most cost-effective.

MICROFILM DUPLICATORS

Microfilm duplicators, which are used to create additional copies of unitized and nonunitized microforms for distribution and security purposes, are of four categories:

- Roll-to-roll duplicators.
- Fiche-to-fiche duplicators.
- Card-to-card duplicators.
- Fiche-to-roll duplicators.

Each of these categories utilize silver, diazo, or vesicular film as either the original or the duplicate (copy) film.

Roll-to-Roll Duplicators

As the name implies, Roll-to-Roll Duplicators are used for the preparation of duplicate continuous lengths of microfilm, which may then be either mounted on reels, in cartridges, or cassettes or be converted to one of the unitized microforms.

The duplicate copies are prepared by the contact process. The original, processed master microfilm in continuous lengths is fed together with an unexposed silver, diazo, or vesicular film into the microfilm duplicator in such a way that the original microfilm is pressed, emulsion-to-emulsion, against the duplicate film. As the two films traverse the microfilm duplicator, they are exposed to a light source that passes through the original master microfilm and exposes the unprocessed copy film.

Fiche-to-Fiche Duplicators

Fiche-to-fiche duplicators, which are similar in concept to the roll-to-roll duplicator, also utilize silver, diazo, or vesicular film as either the original master film or the duplicate copy film. Again the process involves matching an exposed master with an unexposed copy film so that they are in emulsion-to-emulsion contact, and subjecting them to a light source that passes through the original film and creates a duplicate image on the copy film.

Depending upon the specific model, fiche-to-fiche duplicators use one of two types of raw film

- Those that are already cut to standard 148 × 105 mm format.
- 105-mm-wide continuous lengths of film that will later be cut to form 148 × 105 mm microfiche.

Card-to-Card Duplicators

Card-to-card duplicators are used to prepare duplicate copies of microfilm aperture cards. As a rule, the original will be the original silver halide negative microfilm and the duplicate copies will be either diazo or vesicular film that has been mounted in an aperture card.

Card-to-card duplicators are available in a wide range of speeds and capabilities, ranging from the low-cost, relatively slow-output speed models that merely create a duplicate copy from a master, to the more sophisticated models that operate at faster rates of speed and are capable of duplicating not only the microimage, but also the various data that has been keypunched into the master. Finally, at the high-end of the scale, we have totally automated card-to-card duplicators that, in addition to creating a duplicate of both the microimage and the key punched data, can also read a distribution code that has been key punched into the master to automatically control the number of duplicate copies that are required for distribution and security purposes, and to be able to add, delete, or revise data that is keypunched in the master aperture card, without any human intervention, during the time the duplicate aperture cards are being processed.

There is one problem inherent with card-to-card duplicators designed for diazo output. Since, as we discuss in Chapter 4, diazo films are developed using ammonia vapors, care must be taken to assure the card-to-card duplicator is properly vented.

Fiche-to-Roll Duplicators

Fiche-to-roll duplicators are used for the creation of duplicate continuous lengths of 105-mm-wide microfilm from individual microfiche. As such, they form a high-speed method for producing duplicate microfiche in large volumes. Roll-to-fiche duplicators, as the name suggests, produce individual 148 × 105 mm microfiche from continuous lengths of 105-mm-wide microfilm. Each of these duplicators accept either silver, diazo, or vesicular films as either the original master or the duplicate copy films.

Evaluative Criteria

In determining the category and model within that category of microfilm duplicator that will satisfy his organization's requirements, the analyst should ascertain the following:

- The format of the original and the duplicate copies.
- The width and lengths of the original and duplicate films.
- The type films the Duplicator will accept.
- The average daily volume of copies prepared.
- The average number of copy films created from a single original master film.
- The dimensions of the microfilm duplicator.
- Any unique installation and operating requirements, such as the need for venting, and so on.

The determination of the above will very quickly narrow the choice of microfilm duplicators to a single category, then to a specific number of models in a given price range within that category. Finally, these few models should be evaluated in-depth, and if possible tested under actual operating conditions using actual original master microforms, to select the model most suitable to the organization's requirements.

It should be noted that virtually every commercial microfilm processing laboratory also offers a full range of microfilm duplication services, producing copies in every format and on every type of film. An economic analysis, similar to what will be described later in this chapter should be conducted to assure that, from a cost basis, the purchase or lease of a Microfilm Duplicator may be justified. Of course, if there are operational considerations that prompt the acquisition of an in-house microfilm duplication capability (which would be similar to those that would necessitate the acquisition of an in-house microfilm processor), the cost savings will be of lesser importance.

MICROFILM READERS

The Microfilm Reader (Figure 3-10) is a viewing device that enlarges a microimage to readable proportions and projects that image onto a self-contained screen. As such, the microfilm reader is the one indispensable hardware component in any micrographics-based system. All other hardware—cameras, processors, and duplicators—need not be physically present on the user's premises, if he utilizes a commercial service bureau. Microfilm readers, on the other hand, must be available, in sufficient numbers, and in convenient locations, to assure the ability to reference the microimages.

There are literally as many different types of microfilm readers as there are microforms, since there has been a tendency for manufacturers to design readers to meet the express requirements of a specific microform, and often of a specifically encoded or indexed microform. Consequently, microfilm readers on the market today range from inexpensive hand-held varieties as are used for the occasional reading of microfiche, micro-opaques, and microfilm jackets, to large, stationary units designed for the automatic retrieval of specially encoded microforms.

Figure 3-11 is a schematic that is representative of the typical microfilm reader. Although this particular schematic is of a reader that accepts continuous lengths of microfilm mounted in open reels, its method of enlarging a microimage and projecting that enlarged image upon a self-contained viewing screen for subsequent reference is typical of most microfilm readers, in general. There are five basic components to the typical microfilm reader:

• *A transport mechanism,* which may be either manual or motorized.

Figure 3-10   Microfilm reader. Courtesy 3M Company.

- *A light source.*
- *A lens* of suitable magnification to assure the proper enlargement of the microimage to be viewed.
- *One or more mirrors* positioned inside the microfilm reader.
- *A viewing screen* upon which the enlarged microimage will be projected for reference.

Referring to Figure 3-11, let us trace the operation of the microfilm reader to learn how the enlarged image is created and projected for viewing.

The microform is inserted into the reader. In the case of the reader depicted in our illustration, this involves mounting the microfilm reel on the rear spindle, threading the microfilm between the glass flats, and inserting its leading edge into the forward (or take-up) spindle. The electrical power is next turned on, illuminating the light source. The microfilm is then wound onto the take-up spindle by means of the activation of the transport system, which may be either

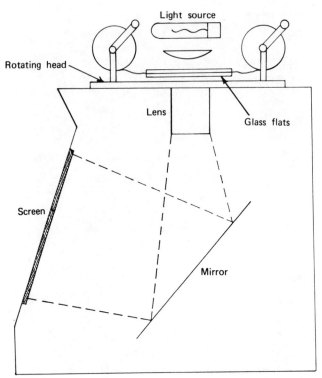

Figure 3-11   Schematic representation of a microfilm reader.

motorized or manually operated. When the desired microimage is positioned directly under the light source, the transport mechanism is stopped. Now, the light source is able to pass, in a beam of light, through the microimage and is projected by the reader's lens onto one or more mirrors that, in turn, reflect the enlarged microimage onto the reader's screen.

The size of the image that appears on the reader's screen is determined by the *magnification ratio* (frequently called the *blowback ratio*) of the lens. Most readers provide a feature that enables the interchangeability of lenses, thus facilitating the reading of microimages filmed at varying reductions.

There are four categories of microfilm readers marketed today:

• **Lap Readers.** Presently available for use with microfiche or microfilm jackets only, these small, readers are ideal for use away from the office.

• **Portable Readers.** Compact readers that operate from either self-contained batteries, electrical power sources, or even the cigarette lighter of an automobile.

• **Desk-Top Readers.**   Compact readers that are designed to fit, unobtrusively, on a standard office desk. These units are generally electrically powered and offer a wide range of magnifications from which to choose.

• **Stationary Units.**   These are large, electrically powered readers, often designed for use with some sort of automated retrieval system.

MICROFILM READER-PRINTERS

Microfilm reader-printers are micrographic hardware that, in addition to producing an enlarged viewing image, have the further ability to make a paper, enlarged reproduction of a given microimage.

A microfilm reader-printer may be best understood if one considers it to be a microfilm reader to which a printing process has been added. For the microfilm reader-printers, in addition to projecting an enlarged microimage on a self-contained screen for subsequent reference, also has the ability to reproduce that microimage in an enlarged, hard copy, or paper, format.

There are four basic categories of microfilm reader-printers, each category obtaining its name from the printing process used to create hard copy reproductions:

• Silver process printers.
• Electrolytic printers.
• Electrostatic printers.
• Diazo printers.

Silver Process Printers

Figure 3-12 is a schematic representation of a reader-printer that uses a wet silver process to create a hard copy reproduction of individual microimages. It will be observed that this reader-printer is basically a reader with two major modifications:

• The mirror upon which the microimage is projected by the lens is hinged so that it can be raised and lowered.
• A simplified film developing unit and silver emulsion printing paper have been added.

The operation of the wet silver process reader-printer is identical to that of the reader, through the projection of an enlarged microimage upon the viewing screen. However, if an enlarged hard copy reproduction of the microimage is

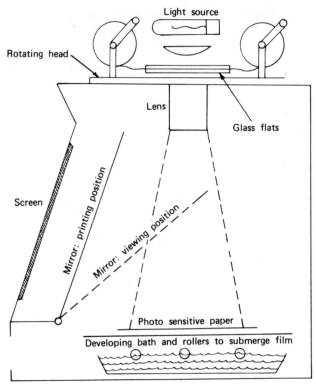

Figure 3-12   Schematic representation of a microfilm reader-printer, wet silver process.

desired, one need only press the "PRINT" button. This raises the hinged mirror from the *viewing* to the *printing position,* so that the projected microimage is now focused on the silver emulsion printing paper positioned directly beneath. After exposure, the printing paper passes through one or a series of chemical baths that develop and fix the image and in less than a minute deliver a nearly dry reproduction of the microimage that is of photographic quality.

A modification of this process is the *Dry Silver Process.* The printing paper used in this process has a developing agent included as part of the silver emulsion coating, and is developed by heat. There are no chemicals or other liquids involved therefore.

Hard copy reproductions produced by reader-printers that use either the wet or dry silver process are generally of better contrast than those obtainable by the other processes available, making them well-suited for applications in which the microimages include solids and half-tones, as well as light or faded lines or data. There are, however, a number of offsetting disadvantages to these processes.

- The *wet silver process* takes longer to produce a hard copy reproduction than the other available processes. There is also a need to replenish the chemicals and liquids used in such processes' developing units.
- The *dry silver process* produces a hard copy reproduction in considerably less time than the wet silver process, and no chemicals or liquids are involved; however, the hard copy reproductions produced by dry silver process reader-printers do tend to begin darkening after several months. As a result, they are unsuitable for use in applications where the reproduction will be retained for longer than 3 years.

Electrolytic Process Printers

The electrolytic (or, as it is also known, the electrochemical) process is another widely used method for preparing enlarged hard copy reproductions of microimages. The printing paper used in this process is of a three-layer construction: a paper backing covered with a thin layer of aluminum that, in turn, is coated with a zinc oxide compound. The resultant printing paper is both light-sensitive and an excellent conductor of electricity.

Microfilm reader-printers that utilize electrolytic process printing have a unique characteristic: two electrodes—one positive and the other negative—form an integral part of their design. One of these electrodes is attached to the tray that holds the unexposed printing papers, the other to the developer bath.

When the print mechanism is activated, a latent image is created in the printing paper as the result of two simultaneous actions:

- The reader-printer's lens projects the enlarged microimage onto the zinc oxide coated surface of the printing paper.
- An electric current passes from the electrode connected to the printing paper tray through the thin aluminum layer of the printing paper.

The exposed printing paper, with the electrode still attached to the aluminum layer, then passes into the developer bath, which is composed of a solution of silver salts that is subjected to a second electric current that is transmitted by means of the second electrode, which is of opposite polarity from the one that is attached to the printing paper.

The electric current is conducted by the silver salts solution to the zinc oxide coating of the printing paper. It then passes through the coating and comes in contact with the aluminum layer, which is being charged by an electrode of opposite polarity. The merging of these two electric currents results in an elctrolytic plating and the development of the latent image.

Electrolytic process printing is a faster method of producing enlarged hard copy reproductions of microimages than is wet silver process printing, but the quality of the resultant reproduction is typically of lesser quality. There is generally less contrast between the background and the data, and a lower density is usually obtained.

Electrostatic Process Printing

Electrostatic (or xerographic) process printing is a fast, dry method of producing enlarged hard copies of microimages on virtually any type of paper.

Electrostatic processes used in reader-printers are of two types: those in which the image is transferred from a selenium drum to the printing paper, and those in which the image is projected directly onto the printing paper.

The first method, which is known as the *Transfer Method,* involves the action of light in forming a latent image composed of varying electrostatic charges on a selenium drum that is contained within the reader-printer. The latent image of static electricity becomes visible when an inky powder that is deposited on the drum, is attracted only to those portions of its surface that represent the image. The image is transferred by contact from the drum to the printing paper and the powdered ink is fixed to the printing paper by heat.

With the *Direct Method,* the image is projected onto a printing paper that has been coated with metallic oxides. The interaction between the projected light and the metallic oxide results in the creation of electrostatic charges that compose the latent image. This image is then developed by the deposit of an inky powder, and is then fixed by heat.

Diazo Process Printing

Diazo process printing is the least used method for producing enlarged hard copy reproductions of microimages. The diazo process involves the projection of the microimage onto a printing paper that has been coated with diazonium salts. The latent image is created during projection on that portion of the printing paper that light does not strike. The exposed printing paper then passes through an alkalide developing bath in which the diazonium salts in the latent image are transformed into an azo dye that becomes the permanent image.

Diazo process printing is relatively slow compared to the alternative processes previously mentioned.

SELECTION CRITERIA—MICROFILM READERS AND READER-PRINTERS

Since there is little standardization among microfilm readers or reader-printers, each must be considered as having its own unique combination of capabilities and limitations. Selection of the reader or reader-printer that will prove "right" or "best" for a given application requires, therefore, a careful definition of the viewing and reproduction requirements of the application under study, and the selection of the reader or reader-printer that most satisfies those requirements, at the least cost.

Determining Which is Needed

The first step is to determine whether a microfilm reader will satisfy the use and reference needs, or if a more costly microfilm reader-printer will be required. As a general rule, one microfilm reader-printer should be available at every location at which microfilmed records will be stored and referenced. This will assure the ability to reproduce enlarged hard copy without undue, and possibly costly, time delays. Once a microfilm reader-printer has been made available any additional viewing equipment may be microfilm readers *unless* 40% or more of the total estimated reference will involve the production of a hard copy enlargement of a microimage. In such instances, an additional reader-printer will be justified.

Once the type of micrographics hardware (reader or reader-printer) that is needed has been determined, one may proceed to the actual evaluation of the scores of such equipment that is available. There are eleven evaluative factors for microfilm readers; sixteen for microfilm reader-printers. Rarely will one individual model reader or reader-printer satisfy all these criteria. Generally, a trade-off will be involved, as one compares the degree that each model satisfies the various selection criteria to first the cost of the equipment, and then to the overall importance of each criteria to the overall system.

Let us now proceed to discuss the first eleven evaluation criteria, which are applicable to both microfilm reader and microfilm reader-printers.

Type Microforms Accepted

As mentioned previously, most microfilm readers and reader-printers are designed for use with one or two different microforms. The first step, therefore, is to determine the various microforms that the reader or reader-printer must accept, and then eliminate from further consideration all such hardware that cannot accpet such microforms.

Further refining the acceptability of various microforms, one must ascertain that the hardware can accept the *specific* characteristics of the microform as well. Therefore, if both 16- and 35-mm film will be involved, only readers and reader-printers that accept microforms of those dimensions will be considered further. Or, if negative microforms are to be used in the applications, hardware incapable of accepting such polarity would be eliminated from further consideration.

By applying this first evaluative criteria, the scores of potential readers and reader-printers will begin to be narrowed down to a more manageable number.

Size of Viewing Screen

The size of the viewing screen must be matched to the user's reference requirements, to assure that the enlarged projected microimage will be readily readable. As a general rule, screens that are at least 8 × 10.5 inches in size will prove adequate for viewing most business records and forms measuring up to legal size (8.5 × 11 inches). However, if the original source document contained small printing, closely spaced lines, or was larger than legal size, a larger viewing screen will likely be required to assure image clarity.

Magnification

Similarly, the size, density, and clarity of the microfilmed data will have a direct effect upon the minimum acceptable magnification ratio. As a rule, greater magnification will be required whenever records are filmed at reduction ratios that exceed 24×, or are drawings, or contain data whose original size was smaller than 9 points.

Image Rotation

Image rotation refers to the ability to rotate the microimage 90° on the viewing screen as well as on any subsequent hard copy reproductions. Image rotation becomes of importance when the microform contains documents that are filmed in both a comic and a cine mode. Unless image rotation is possible, the person referencing the microimage would have to tilt his head to an uncomfortable angle to read the enlarged microimage as it is projected upon the viewing screen, and would in some instances, be unable to produce a paper reproduction of the microimage in a single print.

Need for Subdued Light

Most readers and reader-printers have been expressly designed for operation in a lighted office environment; however, there are still numerous ones that must be operated in subdued light to assure that the image does not wash out on the viewing screen. The ability to operate the reader or reader-printer in a well-lighted area should be the standard, any hardware incapable of yielding a sharp viewing image in such lighting should be considered deficient.

Scanning Ability

Again, this factor applies only to roll film that is mounted on open reels or in cartridges. One must determine if the hardware has the ability to scan the mic-

,roimages at a slow, steady speed. This feature, in the writer's experience, leads to faster retrieval and a decrease in damage to the microfilm that results from rough winding and rewinding.

Internal Coding Systems

There is absolutely no standardization among readers and reader-printers with respect to internal coding systems. Some provide for none, others for code-line indexing, others for the more complex binary coding. The internal coding capabilities of the reader or reader-printer must, naturally, be compatible with that used for the microforms. This is a simple factor to ensure, it merely requires an awareness on the evaluator's part.

Type of Lens

There are two important features that must be considered. First, is interchangeability of lenses to achieve varying magnifications a desirable factor? Secondly, is there a need to be able to zoom in on a given portion of a microimage?

As the general rule, interchangeability of lenses, apart from being economically desirable, is important from an operational standpoint where two or more magnifications are required. The availability of a zoom lens proves valuable where the user must enlarge a portion of the microimage for closer, more detailed examination.

Ease of Loading and Operation

The procedure for loading, operating, and unloading the readers and reader-printers range from slow to very rapid, and from simple to complex and frustrating. This is a key evaluative factor, and a high rating should be assigned to any hardware that is simple to load, operate, and unload. The writer has witnessed a number of applications in which the cost and operational benefits realizable from the micrographics application were rapidly negated by the inability to quickly retrieve and reference the microimage due to the complexity of the hardware.

Drive Mechanism

This factor is critical only if roll or cartridge microforms are involved. If so, determine whether the reader or reader-printer employs a manual or a mechanical drive. If it is anticipated that the hardware will be in use for an average of 15 or more minutes an hour, or that users will line up to use the hardware, then the additional cost of a mechanical drive will be offset by faster, less expensive reference. If not, then the slower, less expensively priced readers and reader-printers that employ manual drives will undoubtedly suffice.

Cost

As with most other manufactured items, the more features that are included in the reader or reader-printer, the greater will be its cost. What to spend for such micrographics hardware involves a value judgement in which the cost of purchase or lease is weighed against the costs of usage. Where a simple, inexpensive reader will suffice if reference is infrequent, coding is simple, and only one microform is involved, such a reader will prove insufficient and ineffective where the opposite conditions prevail.

If the evaluation is to select a suitable reader-printer, the following five evaluative factors must be considered.

Print Process

The use that the various referencers will make of hard copy reproductions of microimages must also be determined to assure that the type of print furnished by the reader-printer will prove satisfactory. The evaluator must ascertain if additional handwritten, typed, or stamped data will be entered on the reproduction, or if it will serve as a master from which additional copies will be produced, and so on. Having made such a determination, actual samples should be obtained of each of the various print processes involved (dry silver, electrolytic, electrostatic, wet silver), and such samples should be tested to see how well they perform or satisfy such secondary processing needs.

Print Paper Supply

Consider too, the type of paper supply used by the reader-printers under consideration. A reader-printer that uses roll paper will generally prove more economical and convenient than one that uses precut sheets.

Size of Print

Reader-printers furnish hard copy reproductions of varying sizes; there is little standardization involved. An examination of the microimages to determine the reduction ratios used, and the reference requirements of the users to ascertain the minimum print size that will satisfy their needs, will indicate which of the reader-printers under consideration will furnish large enough prints. This is perhaps the most important criterion when evaluating reader-printers, for, as experience shows, nothing leads to user dissatisfaction more than the inability to deliver a reproduction of a size satisfactory to the user.

Print Cycle Time

The processing time required to produce a hard copy enlargement of a given microimage becomes important where reference frequency is heavy, since, for all practical purposes, the reader-printer's print cycle time may be considered to be downtime. Where reference is sporadic, the print cycle time becomes less significant.

Cost Per Hard Copy Reproduction

Likewise the cost of producing a single hard copy reproduction from an individual microimage becomes significant as the number of reproductions increases. Individual reproduction costs tend to fall within a very narrow range, so unless the volume is high this factor will prove rather insignificant.

By evaluating the various readers and reader-printers using the above criteria, one may quickly reduce the scores of such hardware to a relative few. The evaluator should then arrange to test the hardware under actual operating conditions to assure that it will perform in a satisfactory manner. Most equipment vendors will be happy to arrange for such a trial period of use.

Maintenance of Readers and Reader-Printers

Microfilm readers and reader-printers contain so few moving parts that their incidence of breakdown is very infrequent. There are, however, a number of simple maintenance procedures that must be routinely performed by users to ensure that the image quality of the enlarged viewing and reproduced image remains satisfactory, and that no damage results to the microforms that are used in this hardware.

On a daily basis, the equipment's lens and glass flats should be lightly brushed using a soft camel's hair brush (which is available at any camera store) to remove accumulated dust. This is important since:

- Any dust on the surfaces of the glass flats could well scratch the silver halide microfilm's emulsion;
- Any dust on the lens or glass flats will be picked up as a part of the enlarged image that is projected on the viewing screen or reproduced on the printing paper. At the least, this dust will appear as stray marks on the enlarged image; at worst, it will obliterate data.

At least weekly, the glass flats, lens, and the interior mirrors should be cleaned using a photographic lens cleaning solution and lens tissues (each of which may be purchased at any camera store). Care should be taken, however, not to inadvertently jar or move the interior mirrors lest the proper positioning of the enlarged microimages on the viewing screen be affected.

Additionally, there are a number of inspections and adjustments that should be performed at least semiannually, or even more frequently if the equipment is continually in use. Such inspections and adjustments are critical to the operation of the readers and reader-printer and require the services of a trained equipment technician. Such adjustments should never be attempted by users; rather the services of the vendor's or manufacturer's field service technicians should be employed to accomplish the following:

- Inspect all drive mechanisms and rollers and adjust as required.
- Inspect and adjust or repair (as necessary) all electrical and electronic circuits.
- Inspect the positioning of all interior mirrors, adjusting as necessary.

MICROFILM JACKET-LOADER

The microfilm jacket-loader (see Figure 3-13) is a specialized item of micrographic hardware that is designed to simplify the loading of one or more microimages into microfilm jackets. Combining into a single item of equipment, the features of a microfilm reader and editor, and a semiautomated insertor, the microfilm jacket loader operates in the following manner:

- The microfilm to be loaded into the microfilm jackets is mounted on the microfilm jacket-loader's supply reel. The loading edge of the microfilm is then wound through the film gates, underneath the reader's lens and inserted into the opening at the end of first microfilm jacket that will be loaded or updated.
- As the microfilm passes under the lens, an enlarged image is projected upon the viewing screen of the microfilm jacket-loader, so that the operator is able to determine at a glance what the next record is to be loaded into a microfilm jacket. The operator can thus determine whether to start a new microfilm jacket or continue loading into the same one.
- The microimages are advanced into the microfilm jacket's channels by mechanical means. The operator advances a crank, or turns a wheel, which in turn moves the microfilm forward and into the channel.
- When a specific channel's capacity has been reached, or when all microimages destined for a particular microfilm jacket have been inserted, another lever activates a blade that cuts the microfilm off squarely and cleanly.
- The foregoing procedure is then repeated, as the microimages are loaded either into the next available channel or microfilm jacket.

The use of the microfilm jacket-loader will significantly reduce the time and costs involved in loading and updating microfilm jackets. Using such hardware, one may anticipate being able to insert up to 1,200 microimages per hour—a

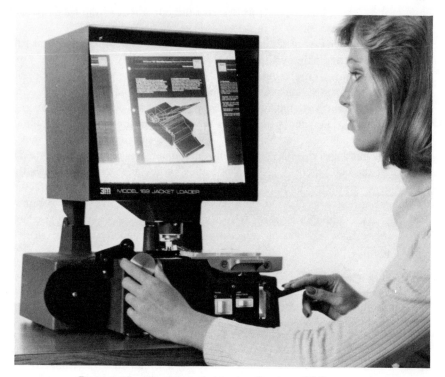

Figure 3-13 Microfilm jacket loader. Courtesy 3M Company.

level that can be easily maintained throughout the day. Manually, a peak production of 350 or 500 microimages in a single hour may realistically be anticipated. Across several hours, or even a whole day, however, the average number of microimages loaded per hour manually will likely drop to a 250 to 350 range.

PURCHASE OF USED MICROGRAPHICS HARDWARE

There is a very active market in used micrographics hardware, affording users the opportunity to purchase readers, cameras, duplicators, and other equipment at savings that average 50% of the original purchase price.

The writer has purchased several items of used micrographics hardware in recent years, and has arranged for several of his consulting clients to do likewise, with generally satisfactory results. There are a number of limitations that must be understood and provided for, however, to assure that one does not fall short of expectations. First, however, let us describe the operations of the used micrographics equipment market.

A vendor purchases or accepts as trade-in on new purchases, used but still serviceable micrographics equipment. He then completely overhauls the used equipment, repairing or replacing all worn and defective components, making all necessary adjustments, cleaning and lubricating all parts, and so forth. The finished product is micrographics hardware that a user may purchase and operate with confidence. The vendor then offers the equipment for sale, utilizing either direct mail or ads in trade magazines. Often the sale will be consummated without the purchaser ever having seen the equipment. The vendor certifies to the buyer that the equipment is in good operating order, and generally allows the purchaser to return the equipment for refund should it fail to perform satisfactorily following delivery. Depending upon the equipment involved, it is common for warranties and guarantees to be issued for motors, moving parts and optics—periods of such guarantee and warranty will generally range between 3 and 12 months.

Evaluative Criteria

In considering the purchase of used, reconditioned micrographics hardware, there are a number of evaluative criteria that should be considered:

• **Warranty or Guarantees Involved.**   Initially, the extent and limitations of any written or implied warrantees and guarantees should be determined.

• **Return Policy.**   Under what circumstances, if any, may the buyer return an item that is not performing satisfactorily? Is such return limited to the repair of the item, or has the buyer the option of obtaining either a satisfactorily-operating replacement or a cash refund?

• **Service.**   Who will service the equipment? Not all used equipment dealers have field repair services, so that a breakdown can lead to undue down-time. Often, the original manufacturer will have a repair facility in your vicinity so that you may make recourse to that repair service.

• **Accessories and Options.**   Can the vendor supply the various accessories and options that you wish for the micrographics hardware you are contemplating purchasing, or must they be purchased from the original manufacturer?

• **Post-Purchase Assistance.**   Will the used equipment vendor provide any training for your personnel in the operation and maintenance of the used equipment?

• **Obsolescence.**   Is the equipment still being offered for sale by the original manufacturer? If not, be sure to ascertain that a reliable source of supply for parts and servicing is readily available and will continue to be so in the forseeable future.

4

Micrographics Technology

An understanding of the types of microimaging films, their properties, the processes by which they create an image, and their archival qualities is important to the design of cost-effective micrographics-based systems.

Three very different types of films are used in source document micrographics and COM applications:

- Silver halide films.
- Diazo films.
- Vesicular films.

Silver halide films are used for both the *original* microform recording media and for the preparation of *duplicate* microforms for distribution and security purposes. The diazo and vesicular films are used as the recording media for *duplicate* microforms.

SILVER HALIDE FILMS

Silver halide microfilm stock, which is used as the original recording media in all source document micrographics and many COM applications, is composed of four layers (Figure 4-1).

- *Layer A* contains finely divided particles of light-sensitive silver bromide or

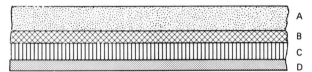

Figure 4-1   Composition of silver halide film.

silver chloride crystals (commonly called *silver halides*) suspended in a gelatin emulsion, which is then coated upon the subbase.

- *Layer B* is a thin binder layer, or *subbase,* that helps the emulsion adhere to the film base (Layer C) through all the chemically active processing application.
- *Layer C* is the film base, a strip of acetate plastic, triacetate, or polyester, that is flexible enough to be transported by rollers through all the steps involved in the microphotography and processing procedures.
- *Layer D* is an opaque undercoating (called an *antihalation undercoating*) that protects the microfilm from the effects of light passing through it and being reflected by the back of either the microfilm or the camera and striking the emulsion from behind (Figure 4-2). Were this to happen, tiny halos would be created around elements of the microimage, thus fogging the microfilm.

How the Microimage is Created

Source Document Microphotography operates on the basis of the absorption and reflectance of light. As the source document is transported in front of the rotary camera's lens, or as it is positioned for filming on the flat bed of a planetary or step-and-repeat camera, it is flooded with light. That portion of the light that falls upon preprinted (fixed) or variable computer, typewriter, or handwritten data will be all or partially absorbed. However, any light that strikes any clear areas of the document (such as the background of the form, letter, report, etc.) will be

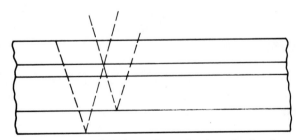

Figure 4-2   How antihalation undercoating impedes fogging.

reflected back through the aperture of the camera's lens and will fall upon the silver halide film's emulsion. When this happens, a change is precipitated in the physical state of the silver halide crystals suspended in the emulsion—they are converted to silver salts.

This reaction of the reflected light on the silver halide crystals is not visible until the microfilm is processed. At that time, when the exposed microfilm is passed through a chemical solution called *the developer,* the silver halide crystals that were converted to silver salts when bombarded by reflected light will be changed to black metallic silver. The amount of metallic silver that will be formed in the emulsion varies directly in proportion to the intensity of the light that is reflected back to the emulsion. The more the reflected light, the more metallic silver will be formed and the darker, or *denser,* will be that portion of the microimage. The converse is also true—the less the intensity of the reflected light, the less the quantity of metallic silver that will be formed and the lighter will be that portion of the microimage. Applying this principle, one sees that greatest clarity of the microimage will be obtained if there is maximum contrast between the preprinted (fixed) and the entered (variable) data, on the one hand, and the paper on which it is printed, typed, or written, on the other. This maximum contrast will be assured if one will adhere to the document design standards that will be presented later in this chapter.

After passing through the developer, there still remains in the emulsion all the light-sensitive silver halide particles that were not bombarded with light reflected from the source document. Therefore, the exposed microfilm must also be treated with a solution that will render these particles impervious to the effects of any light to which the processed microfilm may subsequently be exposed. This is accomplished by moving the developed microfilm through another chemical solution, a "fixer" known as sodium thiosulfate penthydrate (commonly called "hypo"). As this occurs, any silver halide crystals that have not been converted to metallic silver become soluble and are removed from the emulsion, leaving a clear area in their stead. A negative microimage is thereby created.

Hypo is an acidic solution that can prove detrimental to the microfilm; there-fore, it must be thoroughly removed from the processed film. This is ac-complished by transporting the developed and fixed microfilm through a final series of water baths to remove any residual hypo prior to any subsequent format-ting or use. The effects of excessive amounts of residual hypo on microfilm is extremely undesirable, as we shall learn when this subject is discussed in detail later in this chapter.

Figure 4-3 depicts the processed silver halide microform. It should be noted that two significant changes have occurred as the result of its processing:

- The negative microimage has been formed.
- The antihalation undercoating has been removed during processing.

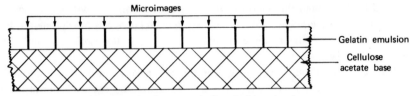

Figure 4-3   Processed silver halide microforms.

Characteristics and Qualities of Silver Halide Films

Silver halide microfilms vary in characteristics and quality in a variety of ways. The task of the micrographics system designer is to assure that the films selected possess the characteristics and qualities necessary to the generation of legible viewing and reproduced images, taking into account the design, format, data entry methods, and reference and use requirements of the source documents.

The micrographics systems designer must evaluate each potential silver halide film in terms of eight characteristics and qualities:

- Type film.
- Polarity.
- Speed.
- Resolving power.
- Granularity.
- Density.
- Color sensitivity.
- Thickness.

After determining the specific requirements of the micrographics system under study, the micrographics systems designer can then refer to the microfilm manufacturers' film specification sheets and rate each potential silver halide microfilm on a scale of from *zero* (the lowest) rating to *ten* (the highest rating) based upon how well the particular film meets the requirements of the source document micrographics system. It then becomes a relatively simple task to select the silver hallide film that has the highest numerical rating, and which appears to most satisfy the reference and use requirements of the micrographics system. A pilot test of that film, or group of films, using the actual source documents and microfilm hardware that would be used in the micrographics system, will indicate if the particular silver halide film will indeed prove satisfactory. Figure 4-4 is an example of such an analysis.

Evaluative factor	Film A		Film B		Film C	
	Manufacturer's Specification	Rating	Manufacturer's Specification	Rating	Manufacturer's Specification	Rating
Polarity	Neg.	10	Neg.	10	Neg.	10
Speed	250	0	64	10	225	2
Resolving power	450	0	670	10	540	4
Granularity	6.0	10	7.5	7	10.5	0
Density	0.10	10	0.12	0	0.10	10
Color sensitivity	Pan	10	Pan	10	Blue	
Thickness	5.4	1	2.7	10	5.5	0
Total rating		41		57		26

*Mfr.'s Specif. = As reported in Manufacturer's Specification Sheet

Figure 4-4   Silver halide microfilm analysis format. Manufacturer's specifications are as reported in the manufacturer's specification sheet.

Type of Film

There are two types of silver halide microfilm:

- Camera film.
- Print film.

As the name suggests, camera films are designed for use as the original micro-graphics recording medium. Print films, on the other hand, are intended for use in the creation of duplicate microforms prepared from the original camera film.

As the first analytical step the micrographics systems designer must determine if the system will require only camera film, or if print film will also be prepared. This initial determination will define the scope of the subsequent analysis.

Polarity

Polarity refers to the type of image that appears on the microfilm:

- *Positive polarity* indicates that the total values of the microimage will be identical with the original source document; whites appear as white, and blacks as black.
- *Negative polarity* indicates that the tonal values will be reversed from the original source document; whites will appear as black, and blacks will appear as white.

Currently, all silver halide camera films used for source document micrographics are of negative polarity, so that this feature does not become a factor in selecting a suitable camera film.

Print films, however, are available with either a positive or a negative polarity. The selection of a print film with a positive or negative polarity will be largely dependent upon the personal preference of the primary users, as much as being determined by the reference and use requirements of the micrographics system itself.

Many users prefer to view a positive microimage. This, according to the author's investigation, appears due more to a sense of familiarity (the users are accustomed to printing, movies, and television in which a positive polarity is viewed), than to any actual difficulties in referencing negative polarity microimages.

There may be valid reasons in specifying either a positive or a negative polarity in a given system, such as the need for a hard copy reproduction that is of a specific polarity. These must be identified and then be taken into consideration in the evaluation of the various silver halide print films available for use.

Speed

Film speeds indicate the sensitivity of the silver halide microfilm's emulsion to light. Film speeds are expressed in a number of different ways, such as ASA ratings, CIS ratings, and DIN ratings. Regardless of the film speeds rating system employed, however, the general rule of thumb is "the higher the rating number, the faster the speed and the greater the sensitivity of the microfilm's emulsion to light."

In evaluating film speed requirements, one must first determine:

- The type of microfilm camera that will be used to film the source documents.

- The quality of the images entered on the source documents.
- The required image quality of the microimage.

High-speed silver halide camera films are required for use in rotary microfilm cameras. The need for such sensitivity is essential since the rotary microfilm camera's exposure time is necessarily brief, and cannot be varied to accommodate light or dark source documents. This shortness of exposure results in the inability to produce a microimage that is always equal in quality to the original source document. Although the microimages produced by rotary cameras will be legible, if the original source document itself is legible, they will also be somewhat grainy in appearance and generally will lack the fineness of detail that is contained in the original source document and that could likely be obtained if the amount of the camera's light intensity were varied.

As the result, the determination must also be made as to the quality of the original source documents and the required quality of the resultant microimages. The poorer the quality of the source document, the less satisfactory will be the image produced on high-speed silver camera film used with rotary microfilm cameras. If the primary purpose of the micrographics system is to record data on film to achieve space recovery or to duplicate records for security and vital records protection purposes, the lack of a truly quality microimage may present little or no problem. If, however, the microimage will be used for active reference or for micropublishing purposes as in the case of Parts Catalogs, or if the original source documents contain data in graphic form, such lack of high quality may prove to be unacceptable. It may then be necessary to either:

- select a slower speed microfilm for use with the rotary microfilm camera, if testing indicates that the use of a slower speed silver halide camera film will produce a microimage of satisfactory quality, or
- film the source documents using a planetary microfilm camera. Such a camera, you will recall, permits the variation of the intensity of the light falling on the source document to adjust for its lightness or darkness.

A third alternative that will prove feasible for future microfilm operations is to redesign the source document to assume that a higher quality microimage is obtained. Such design considerations and guidelines will be presented later in this chapter.

Planetary microfilm cameras, as described in Chapter 3, microfilm stationary source documents one at a time, and can vary their light intensity and reduction ratios to compensate for the quality of the source document. Since it is possible to compensate for lightness and darkness during the actual microfilming operation, it is not essential that high-speed silver halide films be used with planetary microfilm cameras. Slower, less grainy camera films may therefore be used. The result will be microimages whose quality approximates or equals the original source document's.

Resolving Power

Resolving power refers to the ability of the silver halide film's emulsion to record fine detail by reproducing closely spaced information on a microimage, while still maintaining separation between them. Resolving power is expressed in terms of lines per millimeter; the greater the number of lines per millimeter, the greater the resolving power and the higher quality the microimage.

The need for high resolving power is dependent upon the reference and use requirements of the microimage. If the microimage will be used as a micropublishing master or will be employed in an active system in which clarity and accuracy is essential (as in the case of engineering drawings), high resolving power is essential. Conversely, if the application involves ordinary business records, such as invoices, purchase orders, and the like, such clarity will not usually be required, and high resolving power will not be an essential feature of the silver halide microfilm.

Granularity

Granularity is a measure of the coarseness of the grain that might be expected in the microimage. The higher the value, the higher the granularity, and the lesser quality will be both the microimage and any hard copy reproductions made from those microimages.

It should be noted that graininess increases as the microimage is magnified for either direct viewing or for hard copy reproduction purposes; that is, graininess becomes more apparent when a microimage is enlarged 24 times than when it is enlarged 18 times.

Again, the quality requirements of the microimage is the key factor in specifying a satisfactory level of graininess in silver halide microfilms. The greater the need for a true reproduction of the original source document's clarity and legibility, the lower must be the acceptable granularity rating of the silver halide camera or print microfilm.

Density

Density refers to the degree of contrast that exists between the image and the nonimage (or background) areas of a microimage. The greater the density rating, the greater will be the contrast between the dark and light areas of the microimage and the sharper or more legible will be the microimage when it is viewed on a microfilm reader's screen or reproduced as either a hard copy or as another microform.

In those source document applications utilizing a planetary camera, one may vary light intensity and reduction ratios to obtain greater density, or introduce filters between the camera's lens and the source document to obtain better contrast. Although it is true that many rotary microfilm cameras provide adjustments to compensate for the densities or different colored source documents, such variations may not be adequate to achieve the image quality, in terms of density, that is required.

In such situations, the choices are either to:

- use a higher density silver halide microfilm, if tests indicate that such microfilms will work satisfactorily with the user's rotary microfilm cameras; or
- utilize a planetary microfilm camera for the microfilming operations.

In either instance, it is advisable to redesign the original source document in the manner described later in this chapter to increase the contrast between the image and nonimage, or the background, areas.

Color Sensitivity

Silver halide microfilms are of two types:

- *Panchromatic,* capable of recognizing all colors and then reproducing them as varying intensities of gray and black.
- *Particularly sensitive to one color,* such as blue, and relatively insensitive to all other primary colors.

Because of its wider response to color variations, panchromatic microfilm is usually specified for most source document micrographics applications. However, in those applications in which a particular color sensitivity must be maintained, such as instrumentation graphs, the use of a silver halide microfilm with suitable color sensitivity will result in increased image quality.[1]

Thickness

Silver halide microfilms are currently available in two thicknesses:

- *Standard thickness,* ranging from 5.0 to 5.5 mils in thickness. Acetate and triacetate-based films are of this thickness.

- *Thin microfilm,* measuring between 2.4 and 2.7 mils in thickness. Polyester-based films are of this thickness.

Since the thin microfilm occupies less space when mounted on open reels, cartridges, or cassettes, their use greatly increases the amount of microfilm that may be stored in such microforms. It is possible to load 215 linear feet of thin polyester-based microfilm in a single open reel, cartridge, or cassette. Were standard thickness acetate or triacetate-based microfilm used only 100 linear feet of microfilm could be stored on such microforms.

The greater storage capacity per open reel, cartridge, or cassette of such thin microfilms will result in less frequent loading and unloading of cameras, readers, and reader-printers and will lower storage space requirements. These efficiencies may justify the use of such thin-based microfilms in high-volume applications.

Testing

Care should be taken to test those silver halide camera and print microfilms that appear to be the most suitable for the application under study. Actual source documents and the cameras, readers, and reader-printers involved in the application should be utilized in this testing. If the microfilms under consideration pass this "field test," the micrographics system designer may be confident that one very critical aspect of his systems has been provided for—the selection of the silver halide camera and print microfilms.

DIAZO MICROFILMS

Diazo microfilms, although too slow for use as camera films, are used extensively as print films in those applications in which duplicate microforms are required for distribution or security purposes.

Diazo films are composed of a thin plastic layer containing:

- diazonium salts;
- chemical compounds called *couplers,* which when they react with the diazonium salts form dark azo dyes;
- a stabilizing agent that prevents the premature reaction of the diazonium salts and the couplers.

This thin plastic layer is applied to an acetate or polyester film base, and either mounted on open reels or cut to microfiche size. It is then packaged until used.

Creation of a diazo duplicate microform involves the contact printing process described in Chapter 3. The original microfilm is fed together with the unexposed diazo print film into a suitable microfilm duplicator that contains an ultraviolet light source. The ultraviolet light passes through the clear portions of the original microfilm and, falls upon the diazo film beneath, decomposing the diazonium salts in the areas struck by the light.

The exposed diazo film is then developed by being passed through a chamber containing warm ammonia gas. The ammonia gas neutralizes the stabilizing agent, allowing the couplers to chemically react with the diazonium salts. This reaction results in the formation of dark blue or black azo dyes that are insensitive to future exposure to light.

The portions of the diazo film that were struck by the ultraviolet light source during the contact printing, having had the diazonium salts burned out (or decomposed), will, when exposed to the warm ammonia gas, be incapable of creating azo dyes. Rather the couplers remaining in these areas will be turned into colorless compounds, creating clear areas. The resultant microimage will therefore be of the same polarity as the original: negatives will reproduce as negatives, positives as positives.

A recent innovation in diazo microfilm technology has been the development of a reversing diazo microfilm—one that reproduces negative images as positives, and positive images as negatives. Such reversing diazo microfilms were developed in response to, and are therefore well-suited for, use in applications in which a reproduction intermediate is required, as well as those in which reference or use requirements necessitate the availability of a microform that is of opposite polarity to the original.

Diazo microfilms have a number of advantages over silver halide print films:

• **Simplicity of Development.** Diazo films require no darkroom or wet-processing equipment for their development. Diazo developers may be set up and operated in ordinary office areas.

• **Greater Resolution.** Diazo microimages generally possess much greater resolution than silver halide microimages.

• **Less Subject to Damage through Mishandling.** Since diazo microimages are imbedded in the film layers, rather than in an emulsion adhered to the film base, it is less subject to scratching, fingerprints, and other mishandling.

• **Less Subject to Environmental Deterioration.** Diazo films are less subject to the harmful effects of variations in temperature and humidity than are silver halide microfilms.

• **Less Costly.** The costs associated with the creation of duplicate microimages are lower using diazo microfilms than silver halide films.

The major disadvantage of diazo microfilm is its tendency to fade with the passing of time, until its images become too faint to be recognizable. This problem becomes more pronounced as the incidence of the diazo microfilm's use and subsequent exposure to light increases. It must be pointed out, however, that the fading process is generally a long-term one. Many diazo duplicates that have been used for a score or more years show no visible signs of fading. Indeed, some aging tests indicate that diazo copies may retain their image quality for 50 or more years. So in the greater majority of business applications, the problem of fading is relatively minor—the retention period for the microfilmed records will be long past before fading might normally be expected to occur. Still, in those instances where long-term research and reference value exists, and in which the microimages will be retained for many years, the possibility of such fading should be borne in mind, and users should adopt the policy of restricting long-term (over 25 years retention) microforms to silver halide microfilm, or else use the diazo microfilms but adopt the periodic quality inspection techniques that will be described later in this chapter.

VESICULAR FILMS

The third major category of microfilm are the *vesicular* films. Used extensively as copy films in either Source Document Micrographics or COM applications, vesicular films work on the principal of *light scattering,* rather than *light absorption,* as do both silver halide and diazo films.

The vesicular film itself is composed of an ultraviolet, light-sensitive diazonium salt that is dispersed in a thermoplastic resin emulsion. This emulsion, in turn, is coated on a polyester-base carrier—typically Mylar.

Upon exposure to an ultraviolet light source, the vesciular film's emulsion decomposes to form gaseous, volatile substances that occupy molecular size "vesicules" or bubbles in the emulsion. These vesicules, in turn, produce an internal strain within the film stock itself. The exposed vesicular film is then developed by transporting it under a heat source. Heating the veiscular film relaxes the strains within the polyester-base carrier, which in turn, reduces the strains within the emulsion. This allows the gaseous substances in the vesicules to react and form extremely small particles that have light refracting properties; that is, the ability to scatter rather than to transmit light. This feature is the primary difference between vesicular and the other types of microfilm.

Vesicular film may be either reversing (creating a duplicate image that is of the opposite polarity from the original), or nonreversing (creating a duplicate image of the same polarity as the original).

Since vesicular films are exposed by light and developed by heat neither a photographic darkroom with its requisite plumbing nor a vented area to remove noxious gases is required, as is the case when silver halide and diazo films, respectively, are used. On the contrary, most users locate their vesicular duplication equipment right in their office areas or in their computer room.

Of the three types of microfilm mentioned, (silver halide, diazo, and vesicular), vesicular is the fastest and least expensive to produce. As a result, it is being increasingly used for duplicate films intended for distribution and security purposes. For the reasons discussed below, however, many organizations choose to utilize only silver halide films for microfilm-based records—either original or duplicate copies—that will be retained for long periods of time.

ARCHIVAL STABILITY OF VESICULAR AND DIAZO FILMS

Unlike silver halide film, there are no universally recognized and accepted standards for measuring the archival quality of microimages produced on either vesicular or diazo films.

Various studies have indicated that properly developed vesicular films will retain their image quality almost indefinitely. However, there have been numerous reported instances in which the vesicular film was not exposed to an ultraviolet light source for a long enough period of time during the duplication process, and the microimage darkened over the years, causing a lack of clarity, readability, and reproducability. Likewise, there have been reported instances of microimages on diazo film fading over the years, particularly when exposed to light sources for extended periods.

As the result of these experiences, and due to the absence of accepted industry-wide standards of measurement of the archival quality of either vesicular or diazo films, many organizations have stipulated that only silver halide films are to be used for microimages that must be retained for long-term legal, reference, or archival purposes. Such organizations restrict their use of vesicular and diazo films to distribution copies of microforms only.

COLOR MICROFILM

For those applications in which color is necessary or preferred (such as diagrams of complicated electronic circuitry where color-coded wiring is the principal clue to the design, the creation of anatomical records for medical files, and for such micropublications as travel brochures and four-color magazines), color microfilm is obtainable from a variety of microfilm manufacturers.

Available for use in both the creation of original and of duplicate microforms, color microfilm utilizes a complex triple layer emulsion that places a limitation on the film's resolving power (or resolution) that is typically 25% lower than would be obtained were standard silver halide negative film used. Consequently, in those applications in which readability of the microimage of the primary selection criteria and the presence of color is either secondary or a nicety or preserve the format of the original hard copy record, serious consideration should be given to the use of the color microfilm.

Color microfilm requires precision processing and the establishment of and adherence to rigid quality controls, both during processing and afterward during use and storage. This leads most film manufacturers to recommend against in-house processing of exposed color microfilm. They point out that the continuous film processors used for in-house microfilm (described in Chapter 3) are incapable of delivering processed color microfilm of the standards that would be required for the type of reference and use to which color microfilm is intended. As the result they recommend, and every micrographics technician with whom the writer has discussed this subject concurs, that the processing of color microfilm should be left to commercial microfilm processing laboratories.

It should be noted that just as a color photograph reproduced using a standard office copier will yield a black and white copy, so too will color microfilm yield black and white copies when duplicated using standard silver halide negative film, diazo film, or vesicular film or when reproduced in an enlarged hard copy format using a standard reader-printer. Although the viewing image of color microfilm will be in the same colors as the original, any microform or hard copy produced from that color microfilm will be black and white unless special imaging materials and processing is used. This factor too should be borne in mind when evaluating the need for color microfilm.

SAFEGUARDING SILVER HALIDE MICROFILM AGAINST DAMAGE AND DETERIORATION

Silver halide microfilm is composed of potentially unstable materials that are prone to such varied damage and deterioration as the growth of molds and fungus, brittleness, cracking, scratches on its surface, and discoloration, among others. Due to its unique chemical and physical properties, silver halide microfilm is subject to a number of hazards that could adversely effect its image quality and the permanence of the microform itself. Specifically, these hazards include the effects resulting from:

- Improper developing.
- Storage in an area in which atmospheric conditions are uncontrolled and/or acidic gases are present.
- Mishandling, incorrect handling, poorly maintained readers and reader-printers, and the like.

These hazards, it will be noted, are all the result of some improper or inadequate action, which can be prevented, if every user adopts and adheres to a series of specific safeguards and standards. Microfilm, properly developed, stored, and handled will last for over 50 years without noticeable loss of image quality or physical deterioration. If improperly developed, stored, or handled, it will quickly deteriorate. It is incumbent, therefore, that every user understand the potential hazards to which silver halide microforms are subject, and that he take immediate action to preclude the occurance of such hazards.

Improper Developing

As discussed earlier in this chapter, the processing of microfilm involves the passing of the developed film through a solution known as sodium thiosulphate penthydrate (hypo), which fixes the negative microimage and makes it impervious to the effect of light. After the hypo, the microfilm undergoes a series of water baths to remove any hypo from the surfaces of the microfilm. If the greater majority of the hypo is not washed from the surfaces, one of the following undesirable actions may well occur:

• The residual hypo may, after continued exposure to the atmosphere, eventually regenerate a strong enough chemical solution to cause discoloration (in the form of a yellow staining) or fading of the microimages, resulting in an overall loss of image quality and clarity.

• In time, the residual hypo may penetrate the microfilm's emulsion and react with he silver halide therein. This will lead to the development of a rust or brown-colored stain, and a reduction in contrast between the background and data. This reduction in contrast, will, in turn, lead to a loss in image quality and clarity.

The only way to prevent residual hypo on microfilm is to assure the microfilm is properly and completely washed after passing through the fixing bath. A rapid flow of fresh water that is warmer than 60°F is necessary to assure that all excess hypo is removed. It is wise to submit all microfilm to a test to determine the level of residual hypo present on its surfaces. Such a test is a relatively reliable, inexpensive way of reducing the possibility that any hypo will decompose and lead to the deterioration of the microimages. The current American National Standards Institute (ANSI) Standard defines an acceptable level of residual hypo as not eeceeding 0.7 μg thiosulphate ion/cm^2 microfilm, as measured by the Methylene Blue method. Although this standard applies to "Archival" microfilm, the writer's opinion is that it should be regarded as the maximum acceptable level for all microfilm records regardless of whether they will be retained 2 years or 50.

This first hazard is thoroughly controllable. Users must institute a continuing program of assuring proper washing of processed microfilm and assuring residual hypo levels do not exceed ANSI Standards, however, if they are to preclude damage or deterioration from this hazard.

Improper Storage Practices and Storage Environments

As mentioned previously, the very composition of silver halide microfilm is conducive to the development of such deteriorating conditions as mold, fungus, cracking of the emulsion, and the like, that are the result of either improper storage practices or the storage of such microfilm in areas in which the environmental conditions (temperature, humidity, presence of acidic gases) are not conducive to the safe storage of silver halide negative microfilm.

Silver halide negative microfilm should be stored in an area in which the following environmental conditions are maintained:

* *Relative humidity* should remain in the 35 to 50% range.
* *Temperature* should remain between 60 and 72°F.
* The air in the storage area should be free of such acidic gases as sulfur dioxide, hydrogen sulfide, and nitrous oxide.

In addition, the following practices should be implemented to further minimize the possibility of damage or deterioration:

* Do not wrap rubber bands or adhesive tape around rolls of microfilm. Such substances promote the formation and growth of microfilm blemishes.
* Do not store silver halide microforms near diazo or vesicular microforms. These microforms are of different chemistry and their proximity could foster chemical interactions that will lead to damages.
* Likewise, silver halide microforms should not be stored in cardboard boxes or in the general area of cardboard transfiles. The acidity of such containers has resulted in the formation of ''redox blemishes'' on the microforms. It is recommended that microforms be stored in plastic or all metal canisters and containers.
* To protect the microforms against water damage, raise all microfilm storage cabinets, steel shelving, and the like, at least 6 inches off the floor, reducing thereby the likelihood that the microfilm will be submerged in water.
* Drains should also be provided in the floor of the area in which the microfilm is stored to further safeguard against the immersion of the microfilm in water.

- Should silver halide microfilm become immersed in water, be sure that they are not allowed to dry. Place the microfilm into a container that is filled with water and transport them to a commercial microfilm processing laboratory where they can be thoroughly rewashed and properly dried.

Adherence to the above storage and environmental standards will reduce the likelihood that the microfilm will be subjected to the formation of molds, fungus, discoloration, brittleness, and cracking.

Mishandling

Microfilm is extremely susceptable to damage resulting from improper and abusive handling. A single piece of dust or grit imbedded into a glass flat can scratch hundreds of microimages before being noticed. Likewise, improper handling can lead to further damage.

Improper and abusive handling can be reduced by the re-education of users, stressing the following:

- All readers' and reader-printers' glass flats and lens should be cleaned daily (as described in Chapter 3) to eliminate the dust and particles that could scratch the emulsion and damage the microimages.
- Handle all film by its edges as protection against oils on the users' fingers.
- Do not wind the film too tightly. This can cause abrasions, particularly if dust or grit rubs against the emulsion.
- Also, avoid winding the film too loosely. This could lead to warping, since moisture tends to evaporate from loosely wound reels of film.

SPOT-CHECK FOR DETERIORATION AND DAMAGE

Fortunately, it is possible to conduct periodic spot checks of microfilmed records to detect indicators of deterioration and damage. There are six conditions to look for:

- Unusual odors.
- Mildew or mold.
- Discoloration or fading of microimages.
- Buckling of the microforms or fluting of their edges.
- Brittleness.
- Fine powdery substances forming on the microforms.

The presence of any of the above conditions is an indication that an unsafe situation exists and immediate corrective action is required to prevent further damage or deterioration:

- The presence of unusual odors or noticeable discoloration or fading is an indication that excessive residual hypo is present on the microforms.
- Mold, mildew, fungus, buckling, and fluting of the microform points towards excessive moisture in the storage area.
- Brittleness, tested by noticing if the microfilm breaks when bent five or six times without creasing at the fold, indicates excessive dryness in the storage area.
- A powdery white substance on the surfaces of the microforms suggests the presence of harmful acidic gases.

It is the writer's experience that a 10% sampling procedure will preclude the possibility that any of the above six indicators will go unnoticed long enough for irreversible damage or deterioration to result. With this method, every microform is identified by the year it was created, and 10% of each year's microforms are inspected every quarter. If any noticeable signs of damage or deterioration are observed during this sampling, the remaining 90% of that year's microforms are inspected for signs of deterioration and damage. Since, as we will shortly see, microforms are affected by large-scale defects, this sampling procedure will prove both feasible and reliable.

ASSURING SATISFACTORY QUALITY OF MICROFORMS

There is a definite need for the establishment and application of minimum quality standards for microforms in every organization. Certain records, such as engineering drawings and those retained to satisfy the audit and regulatory requirements of certain of the federal and state regulatory agencies, must meet certain prescribed, precisely defined standards with respect to image legibility and reproducibility. Most records that will be converted to a microformat, however, are maintained for their value to the organization, as a source of future reference, or to safeguard the organization against loss due to some future, unpredictable occurrence (such as a fire, flood, or other natural catastrophe, a lawsuit or theft of the records). With such records, there is a need primarily for their capture in a readable, retrievable format. Precision of detail is not of primary concern; neither is the ability to generate four or more copies from the original microform without significant loss in detail. In short, users have little guidance in defining what should constitute "minimum acceptable quality" for such records. In the absence of such guidance, they have generally either: adopted no quality standards at all, relying instead upon the cursory inspection given processed microfilm by commercial laboratories to uncover any defects; or gone to the other extreme and subjected all microforms to the detailed, costly, time-consuming inspection that is afforded records that fall within the purview of such stringent quality control standards as the Department of Defense Specifications for microfilm copies of engineering drawings and other technical documentation created as part of a contract. Either approach is inappropriate. Doing nothing will eventually result in the inability to retrieve data; the latter approach, applying too rigid quality standards to all microfilmed records, regardless of their use or reference requirements, is overly costly.

The writer has long advocated that as part of every micrographics system, the definition of what constitutes minimum acceptable quality be included in the overall procedure, and instructions or guidelines be documented. Persons inspecting processed microfilms following their return from the processing laboratory will then be able to uniformly apply the standard of acceptable minimum quality for the particular records that have been converted to a microformat.

This necessitates an individual appraisal of each micrographic application to determine the following:

- What are the reasons for converting this record to a microformat?
- How long will this record, in its microfilm format, be retained?
- What percentage of these records will ever be needed for future reference?
- Is it possible to use this record if all data is not completely legible: that is, can its contents read in context, thereby obviating the requirement for total legibility of every letter?
- Which information on the original source record represents "essential detail" that must be accurately reproduced in a microfilm format?
- Are the data on the source documents of adequate quality to assure that the enlarged viewing and reproduced image will be readily and accurately readable?
- In the event the microimages were of unsatisfactory quality, could the data be obtained readily and economically from other sources?

The answers to these questions will largely determine the need for stringent standards and, to a great extent, will determine the need for detailed film inspection procedures. Based upon an analysis of these answers it will be possible to assign classifications ranging, on the one hand, from such records as visitor logs, which are: retained in microfilm formats to reduce storage costs, rarely referenced, generally of poor to fair image quality, and could likely be read in context; to, at the other extreme, engineering drawings that: must be accurately read, will serve as the source for additional distribution copies, and are frequently referenced. One may then match the degree and cost of inspection and the standards and costs of quality to the value of the record being converted to microfilm.

Having determined the answers to the above questions, the next step is to identify the crucial features in the microimages; those items of data whose legibility and reproducibility must be assured. This will vary from record to record, based upon which data are considered to be of primary reference and use importance. With a cancelled check, such primary data might be the batch number and the endorsement signature; with a legal contract, it might be all "nonboilerplate" text. It is essential that a level of readability be established and that all film inspectors assure that such levels are met. Although the legibility of microimages cannot be precisely measured, there are four degrees, or levels, of readability that one may readily use:

- **Excellent copy.** All fine detail is reproduced.
- **Legible copy.** The microimage may be read without difficulty, but serifs in printed letters and other fine detail may not be clearly and readily distinguished.
- **Decipherable copy.** Such microimages are readable with difficulty, since letters such as c, e, and o, and numbers such as 6, 8, and 0 may be partly closed or completely filled in.
- **Partly decipherable copy.** At this level, some of the alphabetic and numeric characters may be undecipherable, lines and drawings may be so light that they are hard to distinguish, or the background may contrast so slightly with the data that the data is very nearly or completely unreadable.

Typically, unless there is an overriding standard governing the legibility and reproducibility of the microimage, specifying a quality level of legible copy for statistical and financial records, and those that will be frequently referenced in the future, and decipherable copy for all remaining microimages (basically those having a low rate of reference, containing largely textual data, and for which there are alternate reference sources) will satisfy most organizations' requirements, while substantially reducing the number of rejected microimages, thereby reducing the overall operational costs of the micrographics system.

Inspection Procedures

From a procedural standpoint, the microfilming of source documents using a rotary microfilm camera may be likened to the production of small machine parts—one sets the equipment and feeds in the materials. A few in-line adjustments may be necessary but as a rule the output is produced with little human intervention. Errors that occur on such a "production-line" basis are general rather than individual in nature, that is, they occur in many consecutive microimages, not merely one. It is therefore unlikely that a defect or error that occurs in one microimage will not have occurred in those immediately preceeding as well as those immediately following. This is because most defects and errors that occur in rotary camera-produced microimages are as the result of either phototechnical factors (such as improper exposure levels, faulty film processing, or the use of too extreme a reduction ratio) or mechanical factors (such as distorted and blurred images, streaks and scratches on the microimage, or uneven density within images). Rarely can defects or errors be traced to the operator. Such defects and errors when they do occur are generally the result of missing documents, incorrectly positioned documents or indexing errors.

Since errors and defects, when they do occur, effect numerous, consecutive microimages, rather than individual ones, there is a definite application for inspection on a sampling basis, rather than on a more time-consuming, costly frame-by-frame basis. With this approach, film inspectors would be instructed to review in detail a given percentage of the total microimages, for example, 10%. The inspector would then, in the case of microfilm in continuous 100-foot lengths, inspect only one in every ten or so microimages. If a defect or error is noted, the inspector will work backward from the defective microimages, until a defect-free microimage is reached. The inspector would then work forward from the sampled microimage until the last microimage possessing that defect is detected. The identification of the defective microimages are noted, so that they may be refilmed, and the sampling and inspection continues. We therefore have a system of inspection on a sampling basis until an error or defect is found, and then a 100% inspection to locate all microimages having that defect or containing that error.

Unless one is concerned with assuring excellent copy, this approach should prove satisfactory. Many may argue that this approach may miss the defect that occurs for only one or two frames. That is true. But it is also true, according to tests conducted by the writer, that even a 100% inspection will fail to pick up all errors and defects. The writer included in a group of records to be microfilmed, some with obvious errors and defects. He also programmed in a number of operator errors, such as miscoding, feeding documents upside down, and including obviously incorrect records among those being filmed. The processed microfilm was then delivered for inspection on a frame-by-frame basis. In the first inspection, only 68% of the defects were found. On the second inspection pass, an additional 18% were uncovered. A third and fourth frame-by-frame inspection raised the total to 98% of all known defects. In other words, a 400% inspection was only 98% perfect. For this reason, the practice of inspecting all microimages other than those requiring an excellent copy level of quality is uneconomical, and will rarely be necessary.

DESIGN OF DOCUMENTS THAT WILL BE MICROFILMED

To a great extent, both the rapid, economical, trouble-free operation of microfilm cameras and the production of microfilm images whose quality is as satisfactory as the original hard copy's are dependent upon the design of the documents that will be microfilmed. If the original form, report, correspondence, or other material is designed and prepared taking into account the particular features and requirements of the various micrographics hardware involved, both the speed and economy of the actual microfilm conversion process and the quality of the resultant microimages will be greatly increased.

There are two tasks that the designer of documents to be microfilmed must perform:

- Selecting a paper stock that will facilitate the actual microfilming of the documents.
- Specifying inks, typography, and formats that will assure the production of a legible, reproducible microimage.

Paper Stock

The paper stock upon which the document will be printed must satisfy the following criteria:

- It must be heavy enough to assure that it will lie flat *without shifting or curling at the edges,* as it is either transported through a rotary microfilm camera or positioned for filming on the flat-bed surface of a planetary microfilm camera
- It must fall flat into the rotary microfilm camera's output hopper.

To assure the above, documents that will be microfilmed should be printed on a paper stock whose substance (or "weight") is 12# or more and whose grain runs perpendicular to the leading edge of the document as it will be fed into the rotary camera. If, for example, an 8.5 × 11 inch document will be fed into a rotary microfilm camera with its 11-inch side as the leading edge, a grain running parrallel to the 8.5-inch side should be specified. Grain is always described as being either "long" or "short," in relation to the longest dimension of the document. Therefore, if the grain runs perpendicular to the document's longest dimension, it will be described as having a "short grain." Consequently, grain that runs parallel to the document's longest dimension will be described as "long grain." The direction of the grain is not a factor if the document will be microfilmed using a planetary camera since movement and restacking of the documents are not involved.

Design Considerations

From a design standpoint, there are two objectives:

- To select inks, papers, and carbon papers whose colors and qualities will assure maximum contrast between the document and the fixed and variable data that will be preprinted and entered on it.
- To specify a typography that will provide a readily readable and identifiable image when projected on a reader's viewing screen, reproduced as enlarged hard copy, or duplicated as another microform.

Since a microimage is created by the absorption of light by alpha-numeric, graphic, and other data entered on a document, and the reflection of light by the background areas in which no fixed or variable data appears, a primary objective of document design is assuring that maximum color contrast will exist between the document's paper stock and any preprinted or individually entered variable data. The accomplishment of this objective will assure that the resultant microimage will be legible when enlarged on a microfilm reader's viewing screen, reproduced as hard copy or duplicated in a microfilm format.

Maximum contrast between the document's paper stock and its preprinted and variable data will result if:

- the paper stock is:
  - white in color,
  - smooth and glossy in finish,
  - free of watermarks and screened areas,
  - printed on one side only,
  - relatively free of impurities,
- all data is preprinted or entered using either black ink, black pencil, black typewriter or computer ribbon or black, medium-hard finish carbon paper, as applicable (since all colors other than black will photograph and reproduce as varying shades of gray, and therefore will contrast less than black

with the paper stock's background, their use should be avoided for documents that will be microfilmed);

- the document to be microfilmed is a multipart form, the original copy should be designated as the copy that will be microfilmed. If that is not possible, the copy to be microfilmed should be located as close as is possible to the original copy of the form set, since the legibility of duplicate copies decreases noticeably as one progresses from copy to copy in a multipart form. Often, it will be necessary to redesign a multipart form so that the copy that will be microfilmed is positioned closer to the original copy, thus assuring a higher quality copy for microfilming purposes.

Typography

In specifying typography the document designer has as the primary goal the selection of type fonts and spacing that will result in readily readable and identifiable microimages. This objective will be met by assuring that:

- the type fonts selected will not be reduced, when microfilmed, to the point where the resultant microimages will be indistinguishable when enlarged for reading and reproduction purposes, or when duplicated to create another microform;
- it will be possible to quickly and accurately differentiate between such similarly shaped alphabetic characters as "a", "e," and "o." as well as between such numerals as "6", "8", and "0."

Accomplishment of these objectives will be facilitated by adhering to the following document design guidelines:

- Preprint all fixed data (such as captions, terms, instructions, and the like) using a medium weight Gothic Sans Serif type font that is 10 points or larger in size.
- Allow for the entry of handwritten variable data at the rate of 8 characters per horizontal inch and 4 writing lines per vertical inch.
- Avoid the use of highly stylized and italicized type fonts for either fixed or variable data.
- Provide that all typewriters, word processors, computer printers, and the like, that will be used for the entry of variable data, utilize a type font that is elite size (12 characters to the horizontal inch) or larger.

RECOVERY OF SILVER FROM USED FIXING SOLUTION AND OBSOLETE MICROFORMS

As the price of silver continues to rise on the world's commodity markets, one would be well-advised to evaluate the feasibility of recovering silver from used fixing solutions as well as from obsolete silver halide negative microforms. Silver is one of the precious metals that is never available in adequate supply to meet the demand, and in recent years its demand is far exceeding its supply. As a result, prices paid for mined silver, as well as anything that can be reprocessed into silver, have been steadily rising. According to a recent governmental report, the total quantity of silver that is currently being produced in the United States will satisfy only 25% of our nation's industrial requirements—and this shortfall is at a time when silver production is at an all-time high. The remaining 75%, representing the deficit between domestic production and demand for silver, must be made up either through the imports of foreign-produced silver, or by the reclamation of previously used silver.

Of particular significance to the micrographics user, in this situation, is the fact that both the fixing solution used in the processing of silver halide negative microforms, and the microforms themselves, contain significant amounts of recoverable silver which can be sold.

Recovery of Silver from Used Fixing Solution

As you will recall from our earlier description in this chapter, when exposed silver halide negative film is passed through a fixing bath during processing, the silver halide crystals in the portion of the film that is not bombarded with reflected light dissolve leaving in their place a clear, transparent area. On the other hand, the portion of the microfilm that has been bombarded by reflected light during the filming process has turned black when run through the developer. It is this configuration of opacity and transparency, caused by the blackening and dissolving of silver halide crystals, that results in the formation of the negative microimage. As more and more exposed silver halide microfilm is passed through the fixing solution, more dissolved silver halide crystals accumulate. Finally, the capacity of the fixer is exhausted, and new solution must replace the old to assure proper image quality. The typical exhausted fixing solution will contain between 0.3 and one troy ounce of silver, which can be recovered using relatively simple-to-operate, inexpensive silver recovery equipment that is especially designed for in-house use. As an alternative, used fixing solution may be sold to commercial reclaimers, who will themselves recover the silver from the exhausted fixing solution.

Silver recovery equipment that is used for the recovery of silver from used fixing solution are of three types:

• **Electrolytic Plating.**   Recovery units of this type apply direct electrical current to a series of electrodes that are inserted into the fixing solution. The application of the electrical current causes the silver to leave the solution and accumulate on the electrodes. Periodically, the silver is removed from the electrodes and forwarded for refining.

• **Chemical Precipitation.**   With this type of recovery unit, a catalyst (that is, a chemical that precipitates change) is added to the used fixing solution. The catalyst causes the silver contained in the fixing solution to turn into a sludge, which is later removed, dried, and delivered to a metal refinery where the silver is melted. The use of chemical precipitation silver recovery units involves the presence of rather noxious hydrogen sulfide odors (which smell like rotten eggs), and is a rather dirty operation.

• **Metallic Replacements.**   Silver recovery units of this type operate on the basis of replacing the silver in the fixing solution with another metal, such as zinc or iron, which allows the silver to revert to its solid metallic state. The major disadvantage to metallic replacement-type of silver recovery units is that they recover only the silver, and not the fixing solution, which must be discarded. This type of silver recovery unit is inexpensive, however, accounting for its popularity among small micrographics installations.

Determining the Feasibility of an In-House Recovery Operation

The decision to install an in-house silver recovery from used fixing solution operation or to sell such solution to reclamators, is one that should be made purely on the basis of net dollars realized. If in-house recovery will return greater monies than will sale of the fixing solution to commercial reclamators, then an in-house silver recovery unit should be installed; if not, then plans should be made for the outright sale of the used fixing solution.

Estimate the total yield that may be expected from your organization's used fixing solution by the following calculations:

• Determine the total ounces of silver that will be recovered annually by the following formula:
  • Gallons of solution used monthly × silver content per gallon = total silver content in ounces. (From experience, we can as a rule of thumb assume that every gallon of fixing solution used to process 16- or 35-mm

silver halide negative film will yield approximately 0.5 troy ounce of silver for every 8,000 square inches of microfilm processed.)
For example: 60 gallons × 0.5 oz/gal = 30 oz recovered.

- Next, determine the value of the recovered silver by multiplying the total ounces recovered monthly by the current *bid* price of silver per troy ounce. This latter figure is readily obtainable from the financial section of most newspapers. This will indicate the value of the recovered silver on a monthly basis.
- Finally, multiply the monthly value by 12 to obtain the annual yield.
- This estimated annual yield must be offset by the cost of recovering the silver from the fixing solution. Such costs will vary, based upon the recovery method used. If *chemical precipitation or metallic replacement type silver recovery units* are involved, the following cost factors will be present, and must be determined:

Direct Costs:

- Amortization of the recovery units purchase or rental costs and the purchase or rental costs of all other required accessories, such as shipping and storage containers.
- Labor costs.
- Material costs, such as the cost of the precipitant or metal.
- Transportation costs to the refiner.
- Refining charges.

Indirect Costs:

- Overhead

If an *electrolytic plating silver recovery unit* will be used the following offsetting costs must be determined:

Direct Costs:

- Amortization of the recovery unit's purchase or rental cost.
- Labor costs.
- Transportation costs involved in shipping recovered silver to the refiner.
- Refining charges.

Indirect Costs:

- Overhead.
- Refining loss, (approximately 3% of the gross yield will be lost due to impurities).
- Subtract the costs of recovery from the value of the recovered silver. This

will provide a reasonable estimate of the net monies that might result from the recovery of silver from used fixing solution.

• A price quotation should then be obtained from a commercial reclamator. Indicating what he would pay for a gallon of used fixing solution containing the quantity of dissolved silver estimated earlier. From this figure, deduct the cost of transportation to the reclamator, and multiply by 12. This will indicate the net annual amount anticipated from the sale of used fixing solution.

• By comparing one answer to the other, it becomes a simple matter to identify which will yield a higher income.

Recovery of Silver from Obsolete Microforms

Silver that remains on processed silver halide negative microforms may also be recovered. The recovery process is considerably more complex than that involved in recovery from used fixing solution, and the equipment is more costly. These two facts lead most organizations to sell their obsolete microforms to commercial reclamators for such recovery rather than attempting the task themselves.

Since each pound of obsolete 16-mm silver halide negative microfilm will yield approximately 0.1 troy ounce of silver, most organizations accumulate such obsolete microforms until there is a large number, then sell them to a commercial reclamator. A word of caution—require that the reclamator furnish a Certificate of Destruction, certifying that the microfilmed records have been destroyed without first being copied or revealed to any third party. Since many microforms contain proprietary information, many organizations feel such a Certificate is a worthwhile thing.

5

Micrographics Indexing Techniques

The economical, efficient use of micrographics as an information storage and retrieval media requires the selection of an indexing system that will enable any individual microimage to be accurately and rapidly located without a time-consuming, costly frame-by-frame search. Unless such a system is selected, retrieval will prove as slow, expensive, and cumbersome as locating the proverbial needle in a haystack.

As a rule, the micrographics indexing system selected should be in keeping with the complexity and frequency of reference. It is just as uneconomical and impractical to utilize complex indexing for infrequently referenced records and those filed in sequential order, as it is to use a simple index for records that are constantly references or that may be retrieved based upon a number of variable parameters.

Basically, there are four categories of micrographics indexing systems:

- Serial record indexing.
- Unit record indexing.
- Encoded film indexing.
- Computer-assisted indexing.

SERIAL RECORD INDEXING

Serial record indexing systems are designed to facilitate the retrieval of microimages maintained in *sequential*—alphabetic, numeric, or chronological—*order* in such *nonunitized microforms* as rolls, cartridges, and cassettes. Transaction records, such as cancelled checks, vendor invoices, and time cards, and summary records, such as Accounting Ledgers, are examples of records that are typically filed in sequential order in nonunitized microformats. Included among serial indexing systems are:

- Flash indexing.
- Code line indexing.
- Odometer indexing.
- Image control indexing.

Flash Indexing

Flash indexing is a simple indexing system that subdivides the microfilm into smaller, easier searched segments by inserting blank spaces at predetermined intervals (e.g., every 20 feet). These blank spaces, when viewed in a microfilm reader, appear as white flashes. An alternative method is to insert a numbered target after each segment of the microfilm.

Flash indexing is the simplest, least expensive method of indexing nonunitized microforms, since it requires no special equipment or attachments to either the microfilm camera, the reader, or the reader-printer. It is also the slowest, least accurate, and generally the most expensive method of retrieval. Its use should be restricted to indexing microimages that are basically self-indexing, (i.e., those filed in strict alphabetic, numeric, or chronological sequence, and that will be retrieved using that alphabetic, numeric, or chronological data as the retrieval "key").

Figure 5-1 is a model flash indexing system favored by the author and installed in such organizations as Squibb, the Girard Bank, and Olin Corporation. This system permits the identification of individual microimages contained in a single microfilm roll, cartridge, or cassette, as well as their rapid, accurate retrieval. With this particular flash indexing system each 100 continuous foot length of microfilm, containing between 2,000 and 2,400 images (depending upon whether a 20× or 24× reduction ratio is used), is segmented into five 20 foot lengths, each containing between 400 and 480 images, by means of a clear "flash area." The flash area is created by taping four sheets of blank paper together to form a 34 × 11 inch length. This sheet is then passed through the microfilm camera at the end of every 20 feet of film, and is followed by a flash index target numbered from "1" through "5," and then by the records to be microfilmed (Figure 5-1).

The following procedure is used to create flashed indexed nonutinized microformats:

- Each nonunitized microfilm roll, cartridge, or cassette is assigned a sequential identification number. This number is written on a blank sheet of white letter-sized paper in black magic marker at least 2 inches high. This roll number target is filmed as the first image of the microfilm file.
- A title target is then prepared, also using black magic marker and white paper, describing the title and dates of the records that will follow. This target is filmed as the second microimage.
- Immediately after the title target, the four taped pages are microfilmed, followed by a target reading "FLASH INDEX# 1."
- The individual records are then microfilmed. At the end of the first 20 feet of microfilm, the four taped blank papers are filmed, followed by a target reading "FLASH INDEX # 2."
- The filming of records then continues for the next 20 feet of microfilm, when the third flash area is created.
- At the conclusion of the 100 feet of microfilm, a Certificate of Authenticity is filmed as the last document. The inclusion of such a Certificate of Authenticity is recommended by many legal counsel as a means of increasing the likelihood that the microfilm will be admissable in courts of law and regulatory proceedings. Figure 5-2 is illustrative of the format and content of such a Certificate of Authenticity.

As the actual records microfilming occurs, it is summarized on a simple worksheet, similar to that shown as Figure 5-3. This worksheet is used by the micro-

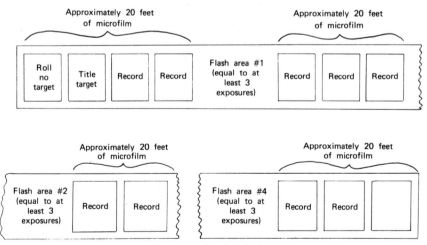

Figure 5-1 Flash indexing.

                    Declaration Of Intent
                           And
                 Certificate Of Authenticity

        This is to certify that the micro-images appearing on this
microfilm file starting with _____
and ending with _____ are accurate
and complete reproductions of records of _____
_____,
delivered to the undersigned by _____
                        of _____,
the legal custodian of such records, who affirmed that such
records were received or made by his organization. Said records
were microfilmed by the undersigned in the normal course of
business, for the purpose of serving in lieu of the original
records in either of the following eventualities:

        First: If the original records should become accidentally
lost or damaged from causes beyond said organization's control.

        Secondly: If said organization should purposely destroy
the original records as part of an established Records Retention
Schedule.

        It is further certified that the photographic process used
for microfilming of the above records was accomplished in a
manner and on microfilm which meets with the requirements of the
National Bureau of Standards for permanent microphoto-graphic
copy.

Date Produced	By (Camera Operator)
Place (City & State)	Camera Operator's Signature

Figure 5-2   Certificate of Authenticity. Courtesy Fenvessy Associates, Inc.

PREPARE IN TRIPLICATE
1. To Records Center
2. To Records Center - Return to Division or Department as receipt
3. Retain in Division or Department

UC 351-40A

MICROFILM ROLL RECORD

DIVISION OR DEPARTMENT	ROLL NUMBER
NAME OF RECORD	

CONTENTS OF ROLL

	FIRST 100 FEET	SECOND 100 FEET (Duo Only)
0		
10		
20		
30		
40		
50		
60		
70		
80		
90		
100		

CAMERA OPERATOR'S CERTIFICATE

I hereby certify that the microphotographs appearing on this roll of film are true copies of the original documents described above.

DATE FILMED	SIGNED

INSPECTION RECORD

I hereby certify that this roll was received and inspected for illegible or missing images.

DEFECTS

If no defects were found, check here ☐

LOCATION	NATURE

DATE INSPECTED	SIGNED	DIVISION OR DEPARTMENT

Figure 5-3   Microfilm worksheet. Courtesy Union Carbide Corporaation.

film camera operator to record the alphabetic, numeric, or chronological filing "key" of the first and last records in each flash indexing area, thus facilitating the preparation of the microform's label (Figure 5-4) utilizing labels furnished by the microfilm's manufacturer.

When flash indexed nonunitized microfilm is to be referenced, the searcher selects the appropriate roll, cartridge, or cassette, using the label as the source. Referring to the indexing data on the label, he determines which flash area is closest to the record he wishes to retrieve. Winding the film rapidly through a suitable reader or reader-printer, he counts the flash areas as they appear on the viewing screen. When the desired flash area is reached, he reduces the winding speed and conducts a detailed, slow scan until the desired record is located. The entire retrieval process, from selection of the microform to the location of the desired microimage should approximate 90 seconds.

Code Line Indexing

Code line indexing is a visual information storage and retrieval technique used in conjunction with specially equipped microfilm cameras, readers, and reader-

INDEX OF CONTENTS

786104 17   MAY.78
EMULSION NO   USE BEFORE
34 7002 6649 6

Figure 5-4   Microfilm label.

printers. With this indexing system, records are microfilmed by a rotary micro-film camera that is equipped with tiny adjustable internal spotlights that are focused on the microfilm as the actual photography occurs. These high-density lights are used to form the indexing code. By setting and repositioning the code line index's positioning controls at various times during microfilming, the micro-film camera operator may select one or a combination of the internal high-density lights to form the code lines that will be recorded, or registered, between each pair of microimages (Figure 5-5).

When the code line indexed microfilm is later viewed in a microfilm reader or reader-printer, these high-density code lines appear as solid dark lines that move horizontally up-and-down the depth of the microfilm, in accordance with the indexing scheme. To permit searchers to accurately and rapidly read these code lines, a scale corresponding to the pattern used in registering the code lines on the microfilm is installed along each vertical side of the viewing screen. By correlat-ing the position of the code bars with this scale, the searcher will be able to quickly and accurately locate any desired microimage. One merely scans the microfilm at high speed, until the position of the code lines along the scale indicates that he is within 10 frames of the desired microimage. During this scanning, the searcher watches the code lines; he does not try to read the indi-vidual microimages. When the code line position that is approximately 10 frames from the desired microimage is reached, the searcher slow scans the following images until the desired microimage is reached.

This indexing method, while slower and more expensive than flash indexing to install, is considerably faster to retrieve; an average retrieval, from selection of the roll or cartridge to the location of the desired microimage, will typically consume 30 to 35 seconds.

The procedure utilized in the creation of code line indexed microforms is as follows:

- The camera operator positions the code line indexing controls at the lowest position. The roll number and title targets are then microfilmed.
- Immediately after, the individual records are microfilmed.
- Depending upon the depth of indexing desired, the position of the code

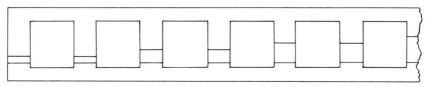

Figure 5-5   Code-line indexing.

lines will be altered after every 10, 20, or 30 records have been micro-filmed.

• The last document microfilmed is the Certificate of Authenticity.

As in the case of flash indexing, a worksheet is prepared as the microfilming occurs, describing the alphabetic, numeric, or chronological filing ''key'' of the first and last record in each roll or cartridge, and a notation of the frequency with which the code lines are repositioned (once every 30 exposures, etc.). Since the typical 100 foot roll may contain as many as 200 different code line positions, it is impractical to attempt to itemize the first and last records in each code line position. As a general rule, therefore, most organizations that utilize code line indexing itemize the first and last records included in every 250 to 400 microim-ages, resulting in between 5 and 12 entries on the roll or cartridge label.

Odometer Indexing

This serial record indexing technique identifies the storage location of each microimage based upon its position in terms of linear location per 100 foot roll. Generally, there are two image locations for every inch of microfilm when a 20× reduction ratio is used, resulting in 2,400 linear positions per 100 foot roll. Each image location is identified by an odometer contained on the microfilm camera that records the linear location of each document. During microfilming, the camera operator records the linear location of the various microimages on the roll. The depth of recording is dependent upon the frequency of the reference to the microimages, as well as the completeness of the records. To illustrate, cancelled checks referenced infrequently, and microfilmed in strict numerical order with few missing numbers, would require less recording of linear locations than would accounts payable invoices files in vendor name order and frequently referenced (Figure 5-6).

Unlike the flash indexing code line indexing systems, the label does not serve as the primary source of reference. Rather, a record locator card or sheet, pre-

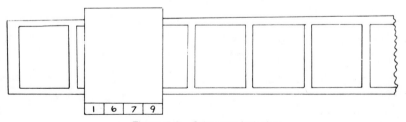

Figure 5-6   Odometer indexing.

pared by the camera operator when the records are microfilmed, serves this purpose.

When reference is required, the searcher first refers to the records locator card or sheet to determine the roll or cartridge containing the desired microimage and the linear location of the microimage on that microform. The searcher then selects and mounts the microform in a suitable reader or reader-printer. The microfilm is then wound through the reader or reader-printer until the desired microimage's linear location number appears on the odometer.

Image Control Indexing

More complete automation of nonunitized microforms storage and retrieval may be accomplished by a technique known as image control indexing. At the time the records are microfilmed, an index code number, generally consisting of the roll number and the linear position of the microimage on the film, is assigned to each document and is generally entered on the face of each document using an automatic numbering head accessory to the rotary microfilm camera. At the same time, an internal high density light positions a mark or "blip" as it is commonly known, directly below the document being microfilmed (Figure 5-7). The index number of each microimage is entered into a supplementary, external index (such as a Uniterm Card or a KWIC Index). When reference is required, this external index is referenced, and the index code number of the desired microimage is noted. The roll or cartridge containing that microimage is selected and inserted into a motorized reader or reader-printer that is linked to an image control keyboard. The index code number of the desired microimage is then entered into the image control keyboard. The keyboard's logic circuitry then scans the microfilm, counting the blips it passes, until the desired index control number is reached. The microimage is then positioned on the viewing screen for reading or reproduction.

The total time for retrieval, from the selection of the roll, cartridge, or magazine to the appearance of the desired image on the viewing screen will typically average under 15 seconds.

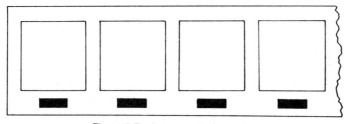

Figure 5-7   Image control indexing.

This indexing technique affords users the opportunity to randomly store records in a nonunitized microformat and still be able to rapidly and accurately retrieve them upon demand.

UNIT RECORD INDEXING

Unit record indexing systems, as the name suggests, are designed to facilitate the retrieval of microimages maintained in either sequential or random order in such *unitized* microforms as microfiche, microfilm jackets, and microfilm aperture cards. Included among unit record indexing systems are the following:

- Self-contained indexing system.
- External index to individual microimage indexing.

As the name implies, *self-contained indexing systems* consist of a unit record microform, in which the complete reference citation, subject, or other descriptive information appears in alpha-numeric characters in a type large enough to be read with the naked eye. Typically, such information appears at the very top of the microform, enabling one to quickly thumb through the entire collection of microforms without having to remove them from their file.

Generally, three types of microforms utilize self-contained indexing systems:

- Microfilm aperture cards.
- Microfiche (both source document and COM generated).
- Microfilm jackets.

Microfilm Aperture Card

Microfilm aperture cards are typically encoded by means of an interpreting keypunch—a specialized data entry device that encodes the indexing data in both an EDP and a human readable form. The data is key punched to allow for EDP input, interpretation, and processing. Additionally, the keypunched holes are sensed (read) by the interpretor, and printed out across the top edge of the card's face, in standard alpha-numerics that are large enough to be read without magnification. This dual language feature—punched holes for EDP and standard alpha-numerics for humans—permits microfilm aperture cardsto be either manually or mechanically filed, retrieved, and updated.

Many micrographics and EDP specialists advise against the mechanical sorting and searching of original microfilm aperture cards. Rather, they recommend that duplicate "slave" decks, without microfilm inserts, be used in mechanical sorting and searching operations to avoid scratching and deterioration of the original microfilm's emulsion.

Data typically indexed in the microfilm aperture card includes an identifying number (such as a drawing number, a part number, an employee number, and the like), and a brief description of the subject or title of the microfilm (such as employee name, project title, etc.) In the case of engineering drawings, it is common practice to include the revision number, and often the revision date in the indexing information. It should be remembered, however, that the indexing capacity of the microfilm aperture card is limited: up to 80 card columns may be encoded for indexing purposes if a "slave" deck is used; up to 53 card columns if the original microfilm aperture card is to be encoded. One would be well advised to request the assistance of their data processing forms designers to assure the inclusion of all the applicable coding techniques that will permit rapid machine processing, and at the same time allow for the inclusion of all possible indexing data required for filing and retrieval. (See Figure 5-8.)

Microfiche

Microfiche, since it is a unitized microform containing numerous related micro-images, poses a dual indexing requirement:

- Indexing the individual microfiche so that it may be quickly and accurately located when reference is required.

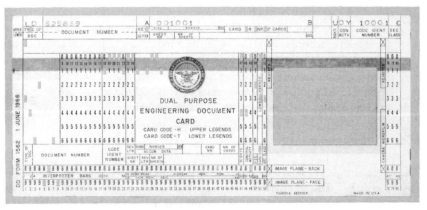

Figure 5-8 Microfilm aperture card indexing.

- Indexing the contents of the individual microfiche so that any microimage may be readily referenced.

Individual microfiche are, in accordance with the American National Standards Institute (ANSI) Standard, produced with a *title block* extending across the top of the microfiche. In this title block, in type large enough to be read by the naked eye, is entered a brief description of the file or records included on the particular microfiche.

Retrieval of individual microfiche may be compared to that involved in conventional filing systems, in which briefly indexed file folders containing numerous related hard copy records are filed. With such systems, one manually scans the individual file folder's tabs until the desired folder is located. The file folder is then manually removed from the file for further searching for the particular information required. With microfiche, one thumbs through the microfiche file, scanning the various title blocks until the desired microfiche is located. When found, the microfiche is removed for further reference using a suitable reader or reader-printer.

A wide variety of indexing techniques may be used to facilitate the location of the microfiche and reference to any of the microimages it contains. From the simplest and least expensive, to the most complex and costly, these include the following:

- Provide no index to the individual microimages. The microfiche is simply placed in a suitable reader or reader-printer and scanned on a frame-by-frame basis until the desired microimage is located. This method is best suited to microfiche that is infrequently referenced, as well as those in which the various microimages are filmed in straight alphabetic, numerical, or chronological sequence.
- Designate each of the microfiche's horizontal rows as the storage location for specific records. Provide no further index to the individual microimages. When it is necessary to reference an individual microimage, the search may be rapidly directed to a small number of microimages located in a predefined portion of the microfiche. This technique will prove faster to reference than will the previous indexing technique.

Microfilm Jackets

The microfilm jacket poses the same external and internal indexing problem as the microfiche. The alternative solutions to resolving these problems are the same, with one exception; the internal indexing used for the microfilm jacket involves the hand-printing or typing of descriptive information on an adhesive strip that is affixed across the top width of the microfilm jacket.

ENCODED FILM INDEXING

Encoded film indexing is a technique that is used in such automated microfilm-based information storage and retrieval systems as Eastman Kodak's MIRACODE and Image System's CARD to provide rapid, accurate, reference and retrieval of individual records that form a part of voluminous records collections. Encoded film indexing combines the microimage of a document and its binary-coded, descriptive indexing on the same segment of the microform. This feature allows one to microfilm documents in a totally random order, but to still be able to rapidly and accurately retrieve them based upon an automated search of the binary-encoded indexing terminology.

Systems utilizing encoded film indexing techniques have three functional components, at least: a special purpose microfilm camera with some means of entering the indexing and converting it to an optical binary code; a search logic, capable of receiving a series of descriptive characteristics, or parameters, and comparing these to the encoded film indexing on the microform; a visual display that can project an enlarged reproduction of any microimage whose indexed characteristics match those of the search onto a screen for viewing and possible production of an enlarged hard copy.

Typically, systems utilizing encoded film indexing provide for the entry, in an optical binary-encoded form, of a variety of identifiers and descriptors relative to the document to be microfilmed, for example, document number, subjects discussed in the record, names of the originator, and addressee or any other information by which the record may be requested. This information is coded at the time of microfilming. Depending upon the system involved, the binary coding may be entered by one of the following:

- The microfilm camera operator who pulls a set of levers on the special-purpose camera to encode the document.
- By a keypunch machine that is linked to the special-purpose camera.
- By a COM unit that creates and positions the optical binary coding on the microform.

When retrieval is required, data descriptive of the desired record are keyed into the system (e.g., document number and subject). The search logic then compares the binary coding of each microimage to the descriptors and parameters indicative of the referencer's requirements. The search continues until a microimage is found whose binary coding matches the search descriptors. That microimage is then projected onto a viewing screen for reference and/or the production of a hard copy enlargement.

Figure 5-9 is an example of one automated microfilm-based information

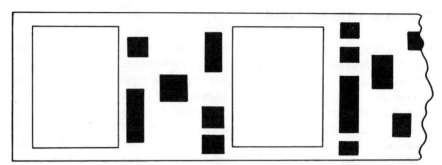

Figure 5-9   Encoded film indexing.

storage and retrieval system's encoded film indexing. No external indexing is required, per se, providing that it is possible to identify the cartridges or other microforms in which a specific group of microimages are stored. This will then enable one to utilize the search logic capacity of the particular system to locate and retrieve any randomly filed microimage contained in that microform.

COMPUTER-BASED INDEXING

The key to rapid, efficient, accurate, low-cost retrieval of any information, other than transaction records that may be indexed and retrieved by such "filing keys" as a transaction number, a vendor's name, or a serial number, has always been in-depth indexing. The more cross-indexing and descriptors that one employs to describe the contents of any record, the more rapidly and accurately will that specific record be found. Therefore, before any system may be considered to be a viable information storage and retrieval system, the ability to index and codify that record in a detailed manner, based upon its contents, is a necessity. In its earlier days, microfilm did not possess this capability. At best, indexing merely directed one to a particular segment of the microform on which the desired record was located; it could not accurately pinpoint the record's location and facilitate its retrieval.

The development of the computer, and more recently the low-cost, easily programmed minicomputer, coupled with such techniques as coordinate index- ing, has brought in-depth indexing capability within the operational and eco- nomic reach of nearly every organization. Additionally, it has enabled microfilm to evolve from an archival to an active information storage and retrieval medium. Today, the computer is used not only to generate microimages via the COM process, but also to generate and update indices to source document and COM- generated microimages.

Use and the writer's experience has proven the following:

- In-depth indexing is essential to the retrieval of all information other than that contained in transaction records that are filed and referenced by a serial number, a name, a date, or some similar "filing key."
- A properly constructed, automated index that permits parameter searching of microimages, and provides for cross-referencing by two or more descriptors or subjects, can reduce retrieval time by over 50%, compared to manual, nonautomated indexing systems. At the same time, the automated index will increase the completeness, accuracy and cost-effectiveness of the retrieval.

Types of Computer-Based Indexing Systems

Computer-based indexing systems are of two types:

- Those in which the computer is used only to identify the microimages that meet the referencer's search criteria and parameters and, having done so, direct the referencer to the microform and location within that microform, in which the microimage will be found.
- Those that, in addition to identifying the applicable microimages, also retrieve and display such microimages for reference and/or reproduction.

Systems of the former type that utilize the computer to maintain, update, and search the index in response to a given set of criteria or parameters, result in a slower, two-phase retrieval. First, the computer-based index must be searched in order to:

- Identify all microimages whose descriptive coding matches the search criteria and parameters;
- determine the identification number of the specific cartridge, microfiche, or other microform that contains each such microimage;
- specify the location of each applicable microimage within the microform. This latter identification, which is not always provided in all computer-assisted-indexing systems, identifies the location of individual microimages in one of two ways:
  - by grid coordinates—the predominant method used in the case of microfilm jackets, microfiche, and ultrafiche;
  - by lineal position within the microform—a method used with cartridge, cassette, or reel microforms.

Once such information is known, the referencer may insert the applicable microform into a suitable reader or reader-printer, locate the applicable microimage, view it on the screen, and, if required, make an enlarged hard copy reproduction of the microimage.

The second method, a concept that is popularly called Automated Document Storage and Retrieval (or ADSTAR), mates the computer and micrographics technologies to effect a system for the storage and retrieval of information that is accessible by any number of parameters. Typically, ADSTAR systems link a minicomputer to a micrographics-based storage system. The minicomputer provides for the rapid, accurate, in-depth indexing and retrieval of information based upon any variety or mix of parameters; the micrographics systems provides high storage density at low cost and an economical means of quickly producing a duplicate copy of any microimage for further reference or distribution purposes. ADSTAR systems that interface the computer and micrographics equipment and technologies will be discussed in greater detail in Chapter 10.

Computer-Based Subject Indexing Methods

The utilization of the computer to maintain and search an index data base enables one to retrieve documents relative to a person, an event, or a thing, based upon a description of its contents. Such retrieval will generally be at a faster rate, producing more accurate results at a lower cost than would be possible were the index manually maintained and searched.

Typically, computer-based subject indexing systems utilize a variety of descriptive terms that define the various subjects, concepts, processes, and other variables discussed in the documents and that could possibly serve as the basis for future reference and retrieval. Each individual application will generally necessitate the development and use of its own unique descriptors, based upon the type of information stored and the probable manner in which retrievals will be requested.

Computer-based subject indexing systems are generally developed following a sampling of:

- the documents maintained in the file and
- present reference requests.

During these samplings, particular attention should be given to ascertaining the various terminology that is used:

- to describe the contents of the documents, if the memorandum, report, and the like, contains a subject heading;

- in the present filing system to classify and store such documents;
- to request the retrieval of the documents from files.

Additional interviews should be conducted with a representative number of users to obtain information regarding their current and projected information storage and retrieval requirements, and their suggestions for making the to-be-developed subject index more attuned to their requirements.

Based upon this information, a subject index may be established. As new documents are received for filing, they will be reviewed and encoded, using these various descriptors. The record itself will then be reduced to a microfilm format and the corresponding descriptors added to the data base (which may, according to the system involved, be either computer-based, output in either a hard copy of a COM format, or be an integral part of the applicable microform).

These descriptors, stored in a computer-processible format, constitute a subject index data base that may be accessed in a variety of ways to fulfill user reference requests. Such computer-assisted subject indexing systems are a primary reason why micrographics has evolved into a cost-effective *active* information storage and retrieval medium.

6

Source Document Micrographics Systems Analysis and Design

A Source Document Micrographics-based information storage and retrieval system comprises a unique combination of records, personnel, and equipment, much of it highly specialized and often unsuited for use in other nonrelated systems. As a result of this uniqueness and lack of interchangeability, it is exceedingly important to assure that any Source Document Micrographics system selected for use will satisfactorily solve the problem at hand economically, efficiently, and effectively. Such assurance is not readily determined, however, since virtually *any* system can be made to work in any given situation.

To illustrate, consider the relatively routine task of transmitting a message between two points. There are likely a dozen ways to accomplish this transmission that come readily to mind—the spoken word, a handwritten letter, braille, telegraphic transmission, smoke signals, semaphore, sign language, and the like. Each of these methods will successfully transmit a message. Some, however, are faster than others; some provide a hard copy record, thereby reducing the possibility of misinterpreting or forgetting the message; others require specialized equipment, skills, or knowledge on the part of the parties involved and so forth.

The same is true of a Source Document Micrographics-based system. Depending upon such factors as operating costs, access time, the need for a hard copy record, the quality of the image, and a host of other variables, any combination of Source Document Micrographics cameras, readers, and reader-printers may be used; as well as personnel with varying skill levels, experience, and knowledge; and microforms and indexing techniques of varying characteristics and limitations.

In its own way, Source Document Micrographics is subject to the same misuses as are computers, and proportionally it is probable that just as many Source Document Micrographics-based systems fall short of their stated goals and objectives as do computerized systems. A thorough, careful, comprehensive systems analysis, therefore, is necessary to evaluate the suitability of microfilm as an information storage and retrieval media, or as a format for communicating or providing a security copy of an essential document, *in the system under study*. Properly conducted, such an analysis will indicate that one or more of the following benefits will result from the Source Document Micrographics-based system:

- Reduced costs.
- Greater efficiencies.
- Increased utilization of available equipment, personnel, and records not presently being fully utilized.

We shall now present a procedure for Source Document Micrographics systems analysis and design—one that will assure the identification and evaluation of those factors that have a direct bearing upon the system's design, as well as upon the selection of the microforms, equipment, and personnel that are required by the resultant system.

PROJECT DEFINITION

Our first task is the definition of the project itself. Project definition involves two phases:

- Definition of the objectives of the Source Document Micrographics-based system.
- Definition of the scope of the proposed system.

Project Definition

To provide guidelines for the project as well as a set of goals against which the new Source Document Micrographics-based system may be evaluated, it is necessary to first define the objectives of the systems change. Objectives will vary from one study to the next, depending upon a particular organization's management philosophy, organization, and procedures. The following are among the most common objectives that prompt the installation of Source Document Micrographics-based systems:

- Space recovery.
- Faster, less costly distribution of printed materials.
- Decreased forms handling.
- Simplified, less costly drafting practices.
- Greater security.
- Reduced clerical expense.
- Faster, more accurate retrieval.
- Lower cost satellite files.

Scope of the project

Determining the scope of the project will necessitate obtaining answers to the following questions:

- Which departments will be affected by the proposed system?
- Who in each of those departments will be affected?
- What records will fall within the scope of the system?
- Which functions will be included in the new system?

The answers to each of these questions will indicate exactly which departments and individuals must be consulted in the systems analysis. Failure to accurately determine who must be consulted could well lead to the system's failure, since the omitted user's requirements may prove to be critical to the success of the system.

DATA GATHERING

Study the Current System

Those developing the new Source Document Micrographics-based system should continue their analysis by studying the existing system, learning everything possible about its strong points and its failings. In doing so, they should interview each Department Head and staff members involved in both the existing and the proposed systems to ascertain how they interact with the system, as well as any unique requirements that they might have. The author's experience indicates that the best way to do this is to actually visit and interview each person. Determine exactly what they do at each step of the present procedure or system. Here the analyst will find that two types of actions are involved:

- What they are supposed to be doing, as defined in formal operating procedures.
- What they are actually doing (the informal procedures) in order to compensate for the inadaquacies of the formal procedures, as well as to satisfy their own unique requirements.

Whenever there is a continuing deviation from the established formal procedure, the analyst should delve more deeply into that operation. By observing the actual operation of the informal procedure, by examining the various records produced and processed during the procedure, the analyst should determine the extent of and reason for the deviation. Frequently, it will be found that the deviation results from a reference or use requirement that is not being met or satisfied by the formal procedure—a requirement that should be provided for in any proposed new system. To facilitate further analysis, both the formal and the informal procedure should be flow charted.

During interviews, the analyst should also determine if any of the individuals are prejudiced either for or against micrographic systems. Uncovering such prejudices early in the analysis phase of the study could indicate potential trouble spots or areas of support, and point the way towards slanting the systems presentation to overcome or reinforce these prejudices.

Determine Reference Characteristics

The analyst should also gather data that will indicate the following:

- The actual or anticipated frequency of reference to the various records that form a part of the Source Document Micrographics-based system.
- The type reference that is normally required; that is, does the referencer merely look up and verify data; does he make notations on the records; does he require a copy of it for subsequent reference?
- The speed of response that is normally required. Does the referencer require immediate access to the information or can he or she wait for a short period of time to receive an answer?

This information will help determine which microforms, indexing systems, and micrographics hardware will be required.

Determine Space and Equipment Requirements

Next, the analyst must physically examine the records that are being considered as candidates for conversion to a microfilm format in order to determine the following:

- The type of filing equipment the records now occupy in both active office and inactive records storage areas.
- The annual rental costs of the floor space occupied by such records.
- The replacement value of the filing equipment and supplies located in both active office and inactive records storage areas that are devoted to the storage of such records.

Determine Existing Records Retention Periods

The analyst should then determine the length of time that each record involved in the proposed Source Document Micrographics-based system is currently being retained. This information is necessary if the analyst is to make a valid comparison between the costs of maintaining the records in a hard copy and in a microform format.

Determine Which Governmental Agencies Mandate the Retention of the Records under Study

To ascertain if there are any legal or regulatory restrictions that would either prohibit or limit the use of a microform as the *sole* storage media for records that are required for audit by the various federal, state, or municipal regulatory agencies or the introduction of the microforms as primary evidence in courts of law or in legal proceedings, one must first identify the various regulatory agencies to whose jurisdiction the analyst's organization is subject.

This can be most expeditiously accomplished by contacting the heads of the below-listed departments and asking each of them:

"To which federal, state, and municipal agencies does our organization submit periodic reports? Also, which federal, state, and municipal governmental agencies have the right to audit our organization's records, and, finally, which agencies have ever obtained a subpoena for the submission of our records?"

The writer suggests that the above questions be directed to the heads of the following department's in the analyst's organization:

- Tax Department.
- Industrial Relations Department.

- Comptroller.
- Legal Counsel.
- Corporate Secretary.
- Operations.

Obtain Samples of Documents Involved

Finally, samples of each document involved in the application under study should be obtained. These samples will prove necessary in the next phase of the study.

DATA ANALYSIS AND SYSTEMS DESIGN PHASES

Gathering of the basic information and samples described above will facilitate the next two phases of the Micrographics Systems Design—the evaluation of the feasibility of conversion to a micrographics-based system and, if applicable, the design of such a system.

Evaluation of Documents

The first task is to assure that the format and physical condition of the various forms, reports, correspondence, and other documents involved lend themselves to rapid, efficient, trouble-free microfilm conversion.

Examine each of the various documents involved, and note such factors as:

- their overall condition;
- the legibility of both preprinted and entered data;
- the size and style of type font used;
- the color of the documents.

As a rule, unless documents to be microfilmed are of a uniformly legible quality, and in good physical condition (free of unrepaired rips and tears; clean with no dirt or smudges) one risks that the resultant microimage will be of less than satisfactory quality. It must be remembered that converting a document to a microfilm format will not improve its legibility. Indeed, since there is a slight loss in image quality as the result of the microfilming process, an original document that is fuzzy or illegible will yield an even fuzzier, more illegible microimage. Consequently, the decision to microfilm a given document will necessitate changes in the record itself. To illustrate, it may prove necessary to change the color of the paper a form is printed on from blue or salmon to white, in order to provide greater contrast with the black typewritten entries. It might also be necessary to reposition the copy of a multipart form set that will be microfilmed, moving it to the original or first carbon copy, affording a more legible source document.

Significant Color Entries

As described in Chapter 4, standard black-and-white microphotography produces a microimage that renders all colors as either black or varying shades of gray. Therefore, any significant colors that are included in the original document will prove difficult to recognize in the microimage, necessitating the use of an alternate coding method. For example, the use of red ink to denote credits in accounting records will prove unsatisfactory if the records were converted to a microfilm format. The use of such alternative coding as underlining the red-ink entries, placing an asterisk to their immediate right, or even enclosing the entire amount in parentheses, are typical of the alternatives that could be used to circumvent the inability to capture the significant colors. Likewise, color coding used to indicate different circuits in engineering drawings, as well as colors used in various charts, graphs, and clinical records will also require the development and use of alternatives to color coding if they are to be accurately referenced following their conversion to a microfilm format. A typical approach would be, in the case of color coded electrical wiring diagrams, to add numbers to the diagram indicating the colored wires, prior to the microfilming. Thus, a green wire would be encoded "1," a white one "2," and so forth.

Presence of Staples and Other Fasteners

Increased costs will be incurred if the documents to be microfilmed are stapled, paper clipped, or fastened in any other manner. Such fastenings will have to be removed prior to the actual microfilming—a time-consuming, costly procedure. As a general rule, it is wise to prohibit, or at the minimum severely restrict the use of such fastenings as staples with documents that will be microfilmed.

Determine Legal and Regulatory Acceptability

If it is determined that the format, physical condition, and quality of the documents under study will economically yield, or may be revised to yield, a satisfactory microimage, the next task is to ascertain if there are any legal or regulatory requirements that will impact upon the proposed micrographics-based system. In this determination, the following must be examined:

- Federal and state laws relating to the admissability of microfilm copies of business records as primary evidence in courts of law and in other legal proceedings.

- The regulations of the various federal, state, and local governmental regulatory agencies to whose jurisdiction the individual organization is subject, concerning the conditions, if any, that must be satisfied in order to retain records in a microfilm format rather than as hard copy.

Generally microfilm is fully acceptable as a records storage media, insofar as its introduction as primary evidence is concerned, in both federal and state courts of law, as well as in legal proceedings within such jurisdictions. In 1951, the Congress enacted the Uniform Photographic Copies of Business Records as Evidence Act, which specified that, microfilm copies may be substituted for original hard copy records. Since that time, 49 of the 50 states have enacted their own versions of the Federal UPC Act (as the 1951 Act is commonly known), which permits the substitution of microfilm for hard copy records within the state courts. Such Acts are the basis for programs in which records are microfilm and subsequently destroyed. The only state yet to adopt a version of the Federal UPC is Louisiana, and the only territory is Puerto Rico. Yet, in even these jurisdictions, courts acting under the Best Evidence Rule have admitted microfilm in evidence when the original records were not available.

In ruling on the admissability of microfilmed records as evidence, many legal counsel suggest that a Declaration of Intent and Certificate of authenticity be filmed as the final record on each roll of film (or microfiche) as a means of establishing that the microfilming was done in the normal course of business and that the records were filmed in their entirety with no alterations or ommissions other than those noted being made at the time of filming. Figure 6-1 is an example of such a form suitable for use with continuous microfilm; Figure 6-2 with microfiche. Still, there are problems.

Many regulatory agencies, although generally accepting the concept that microfilm is acceptable as a records storage media, impose requirements that must be satisfied before microfilm can be substituted for hard copy records.

The Civil Aeronautics Board, for example, specifies that airline carriers must create and retain certain records in their original hard copy form for at least one year before microfilm may be substituted; and in the case of specific records of a long-term audit or statistical value, microfilm may never be the sole records storage media.

The Securities and Exchange Commission, on the other hand, permits stockbrokers and other that they regulate, to substitute microfilm for hard copy records at any time. They do stipulate, however, that if the microfilm is prepared by the computer-output-microfilm (COM) process (the process by which microfilm is created directly from digitized data without any paper involved), a dupli-

Declaration Of Intent
And
Certificate Of Authenticity

     This is to certify that the micro-images appearing on this microfilm file starting with _____ and ending with _____ are accurate and complete reproductions of records of _____, delivered to the undersigned by _____ of _____, the legal custodian of such records, who affirmed that such records were received or made by his organization. Said records were microfilmed by the undersigned in the normal course of business, for the purpose of serving in lieu of the original records in either of the following eventualities:

     First: If the original records should become accidentally lost or damaged from causes beyond said organization's control.

     Secondly: If said organization should purposely destroy the original records as part of an established Records Retention Schedule.

     It is further certified that the photographic process used for microfilming of the above records was accomplished in a manner and on microfilm which meets with the requirements of the National Bureau of Standards for permanent microphoto-graphic copy.

Date Produced	By (Camera Operator)
Place (City & State)	Camera Operator's Signature

Figure 6-1 Declaration of Intent and Certificate of Authenticity, continuous microforms. Courtesy Fenvessy Associates, Inc.

Microfiche/Microjacket Certification Log

This is to certify that the below described records were received
and microfilmed by the undersigned in their entirety, and after
inspection, inserted in microfilm acetate jackets.

Date:	Camera Operator:	
Records Microfilmed	Alphabetic, Date Or Numeric Range	Delivered To

Figure 6-2   Certificate log, microfiche/microjackets. Courtesy Fenvessy Associates, Inc.

cate of the COM-generated microfilm be stored in a location separate from the original microfilm.

The Internal Revenue Service, in its recently issued Revenue Procedure 76-43, has ruled that microfilm may be used for the storage of both detail and summary accounting records (formerly, only detail records could be maintained as microfilm during the period of audit) provided:

- the film is suitably indexed;
- the image quality of both the original microfilm and reproductions meets certain minimum standards;
- a reader-printer is made available for the IRS's use when examining such microfilm records.

Various state regulatory agencies have enacted similar requirements concerning the use of microfilm as the sole records retention media. Certain state banking departments require that banks retain specified records in their original hard copy form for at least 2 years before microfilm may be substituted. A number of state taxation departments have chosen not to follow the IRS' position as outlined in Revenue Procedure 76-43 and still require that summary accounting records, such as General Ledgers, be retained in their hard copy format.

No broad statement can be made about which records can or cannot be freely maintained in a microfilm format rather than as hard copy. Each organization's unique position must be evaluated, taking into account the requirements of each regulatory agency to whose jurisdiction it is subject. A good source of federal agencies' requirements is the Federal Register's "Guide to Records Retention Requirements," available from the Superintendent of Documents-Government Printing Office, Washington, D.C. This annually-updated, inexpensive publication delineates the records retention requirements of every federal agency, and also contains a valuable index indicating, by various industry descriptors, the regulations to which the organizations in that field of endeavor are subject.

As a rule, there are no compendiums of the records retention requirements of the various state or local regulatory agencies. With the exception of the state banking departments, the state taxation departments, the state public utilities commissions, and the state insurance departments, it is unlikely that any state or local regulatory agency will have specified any conditions, limitations, or restrictions on the use of microfilm. In the absence of a readily referenced source, however, it is advisable to contact the Audit or Compliance Divisions of the specific regulatory agencies, to whose jurisdiction the organization is subject, to learn if there are any restrictions concerning the use of microfilm.

It should also be noted that the individual regulatory agency, as in the case of the federal IRS, may require that users desiring to install microfilm-based recordkeeping systems in lieu of hard copy, must first obtain that agency's written approval.

The use of microfilm for records required by regulatory agencies necessitates the exercise of "due diligence" in the storage of the microfilm to assure its long-term preservation. This concept of due diligence means that an organization will make all efforts to assure the safe preservation of microfilms by taking whatever steps a "prudent man" would. For example, recognizing the suscepta-bility of microfilm to damage and deterioration (as described in Chapter 4), due diligence requires that steps be instituted and continually administered to minimize the possibility of such damage and deterioration. At a minimum the following precautions should be taken:

- Microfilm should be subjected to tests after processing to assure that it is acceptable for archival storage.
- It should be stored in an area in which temperature and humidity are within the ranges specified for safe storage and in which no harmful gases are present.
- Periodic samplings should be taken of each year's accumulation of micro-film while in storage to assure that no deterioration is occurring. An annual sampling of 20% of each year's accumulation of microfilm should suffice.
- Any deterioration noted during this sampling should be treated promptly while there is still time to restore the microfilm to a usable condition.

Regulatory agencies do not require the creation and maintenance of duplicate microfilms unless they specifically say so in their regulations and administrative issuances and procedures. Unlike due diligence, where the use of microfilm implies the establishment of minimum environmental and quality control stan-dards, the necessity of providing a security microfilm has not been mandated by most federal, state, or local regulatory agencies.

Acceptability of Microfilm in Foreign Jurisdictions

With few exceptions, microfilm has not yet been ruled to be legally admissible as primary evidence in foreign nations' courts of law. This is largely due to the unfamiliarity of trial judges with microfilm's technology and reliability. Multina-tional companies contemplating the use of microfilm as a substitute for hard copy should first ascertain microfilm's legal status in that country. It will be found that four basic conditions exist:

- Microfilm is acceptable under the foreign nation's version of the UPC Act. This situation prevails in:
  - Australia

- Canada
- England
- Japan
- Scotland
- No UPC Act has been enacted, but petition to the regulatory body involved will generally result in permission to retain specific records in a microfilm format rather than as hard copies. Included in this category are:
  - El Salvador
  - Federal Republic of Germany
  - Italy
- Following acceptance by the trial judge, microfilm may be introduced as either *secondary* or *indirect* evidence. This is microfilm's status in:
  - Belgium
  - Denmark
  - Finland
  - France
  - India
  - Poland
  - Romania
  - Spain
  - Sweden
  - Yugoslavia
- Microfilm records cannot be admitted as legal evidence. Nations in which this is the case include:
  - Argentina
  - Bulgaria
  - Union of Soviet Socialist Republics

It is recommended that any multinational organization that is contemplating the use of microfilm as a substitute for hard copy first contact its legal counsel and independent auditors to assure that the proposed microfilm conversion will not adversely effect the admissibility of the records in either courts of law, legal proceedings, or regulatory actions.

Selection of a Microform

The reference and use requirements that were defined during the data-gathering phase should now be analyzed to determine which microform will prove the most satisfactory. Initially, the decision should be whether to specify a unitized or a nonunitized microform. Since it is more time-consuming and costly to prepare unitized microforms than nonunitized, one must ensure that the reference and use requirements justify the additional expense. As a general rule, unitized microforms should be considered whenever groups of records are referenced as a unit, as in the case of project files, personnel files, and the like; or when frequent updating is required, as in the case of active medical files. Nonunitized microforms, on the other hand, are best suited for use with records that do not require updating and that each record a single transaction. Journal vouchers are examples of records that are well suited for nonunitized microforms.

Having determined that a unitized or a nonunitized microform will prove most satisfactory, one may then determine which *specific* microform will be most feasible. It is recommended that the various advantages and limitations of each type of microform, as described in Chapter 3, be reviewed to determine which will most successfully satisfy such reference and use requirements.

Specify Indexing Methods

The reference and use requirements, as defined during data gathering, will suggest the indexing system that will prove most applicable. In the preceding chapter, each of the various indexing methods, from the rudimentary flash indexing to the complex computer-assisted indexing were described. The guidelines and recommendations for use that are contained in that chapter should be reviewed to determine which will most closely satisfy, at the lowest cost, these reference and use requirements.

Specify Micrographics Hardware

The various data gathered during the initial phase of the micrographics systems design will, to a great extent, point the way to the selection or specification of micrographics hardware. In the selection of suitable microfilm cameras, for example, such variables as the size of the documents to be microfilmed, their format (bound or unbound), the color of both the papers on which the documents are printed and in which data is entered, will quickly narrow the selection process to either a rotary or a planetary camera. By then evaluating such additional reference and use requirements as the microform desired, the indexing methods to be included in that microform, the desired reduction ratio, and the minimum resolution acceptable, among others, one may quickly hone in on a microfilm camera that will successfully provide for these factors as well.

With respect to readers and reader-printers, the same evaluation technique may be applied, contrasting the estimated reference and use characteristics determined or developed during the first phase of the study to the individual readers' and reader-printers' features and capabilities.

In Chapter 3 specific evaluative criteria were presented for selecting various types of equipment. These should be applied here to facilitate the selection of such hardware as is required in the new micrographics-based system.

Determine Who Will Handle the Conversion

There are two approaches that may be taken in the conversion of hard copy to microfilm:

- Utilize in-house personnel and equipment.
- Utilize a commercial microfilm service bureau's personnel and equipment.

During the data-gathering phase, various data relating to the documents under study were obtained; their format, physical specification, and general condition was determined, and current and projected reference and use requirements were identified. This data will prove valuable in the determination of who should handle the microfilm conversion. As the initial step, one should determine if any physical characteristic of the documents under study preclude their being microfilmed using rotary microfilm cameras.

As a rule, unless the smallest dimension of a given record exceeds 12 inches, or the record is bound, or extremely high resolution is required, all records should be filmable using rotary microfilm cameras that are equipped with a suitable indexing capability. Records that exceed these criteria can best be filmed using a planetary microfilm camera; equipment that is often not found in the small-to-medium sized, in-house micrographics unit.

Generally, unless an organization can utilize its microfilm cameras at least 40% of the time, it is not advisable to go to the expense of setting up or expanding its in-house microfilm conversion capability. The organization should rely instead upon a commercial microfilm service bureau. As a rule, such commercial service bureaus will provide an effective, economical alternative to an in-house micrographics capability for those organizations initially implementing micrographics programs as well as for small to low medium users of microfilm, since there will be no capital investments for cameras, processors, and duplicators, equipment that often accounts for the majority of hardware purchase dollars. There would be no need for trained micrographics technicians, such as camera operators, no costly, wasteful underutilization of hardware. The output one may anticipate receiving from a commercial service bureau will typically be of quality that is superior to that which a fledgling in-house micrographics facility could reasonably be expected to produce. Of course, as an organization becomes more deeply involved in micrographics, and as increasing numbers of new applications are implemented, the economic evaluation of in-house versus commercial service bureau microfilming will likely swing in favor of the establishment of an in-house capability. Continued cost comparisons should be made in this regard.

Prepare Preliminary Flow Charts and Systems Specifications

At this time, the various factors are available that will enable one to prepare flow charts descriptive of the new micrographics-based system, as well as a general specification that will delineate such technical factors as the type of microforms involved, indexing methods to be utilized, minimum quality standards relative to the microimage, type, number, and distribution of duplicate microforms, and the like. These preliminary flow charts and specifications will serve a two-fold purpose:

- To facilitate review of the proposed new system by users, management, and those vendors whose products or services are involved in the new system.
- As a training tool for use in implementing and auditing the new microfilm-based application.

Prepare Cost Estimates

Using the data gathered during the initial phase of the study, and referring to the preliminary flow charts and systems specifications, the proposed micrographics-based application should next be costed-out.

Costs may be estimated by either the pilot microfilming project method or by applying established standards to the various factors involved in the microfilming project.

The former approach, in which the microfilming conversion is actually performed using a few thousand documents as the basis for developing production times and costs, is the most satisfactory of the two methods. With the pilot microfilming project approach, both labor and material requirements can be reliably projected. The sample microform containing microimages of "live" records will provide an opportunity to test the suitability of the indexing, reduction ratio, resolution, and other similar features. Any problems resulting from the need to provide special indexing, or a different arrangement of the documents on the microfilm, which may have been overlooked in the planning, will usually become obvious during such a representative pilot project. It should be pointed out, however, that pilot projects are costly, and are feasible, as a rule, only when the organization has an in-house capability. It generally becomes far too expensive to lease suitable micrographics cameras, and hire camera operators to conduct such a test. Also, the costs involved in contracting with a commercial micrographics service bureau for such a study would also be too expensive. If the organization lacks an in-house micrographics conversion capacity, the second approach, the application of standards, will provide an estimate that, while less reliable than is obtained using the pilot project method, will present a representative picture of the overall systems costs.

With this method, the following standards would be used:

- The average 100 foot roll of microfilm will contain 1,800 microimages of letter-sized documents when filmed at 18X, 2,000 when filmed at 20X, and 2,400 when filmed at 24X. Therefore, a typical four drawer letter sized file cabinet containing 10,000 documents would require 4.2 rolls of microfilm if filmed at 24X; 5 rolls if filmed at 20X and 6.3 rolls if microfilmed at an 18X reduction ratio.
- The average microfilm camera operator can microfilm approximately 650 letter-sized documents an hour if using a rotary camera not equipped with an automatic feed and about 500 documents per hour using a planetary camera and documents that are batched by size and reduction ratios. If the rotary camera is equipped with an automatic feed, an average production of 1,000 to 1,100 documents per hour may realistically be anticipated.
- The typical time required to prepare documents for microfilming (preparing targets, arranging all records in the same direction, removing staples, and the like), is 30 minutes per 2,000 documents. If the examination of the files conducted during the data-gathering phase indicates that unusually large numbers of staples or paper clips must be removed, or that numerous records must be repaired before filming, or that color coding must be noted on the document, this production figure should be reduced by 10 to 40%, as the analyst best judges.
- If the processed microfilm will be inspected on a 20% sampling basis, allow 30 minutes of film inspection labor per 100 foot roll of microfilm. This figure will have to be adjusted accordingly if more or less frequent sampling is to be conducted.

Each of the above results will be converted to a dollar figure in order to estimate the *direct costs* of the microfilm conversion, for example, assume that we are interested in estimating the direct costs involved in converting 25,000 letter sized documents to roll microfilm, at a 24X reduction ratio. The documents are in generally good condition and will be inspected on a 20% sampling basis. The technician who will do the preparation, microfilming, and inspection makes $5.50 per hour.

Microfilm	10.25 rolls @ $6.50/roll, including process- ing	$ 66.63
Processing labor	6 hours @ $5.50/hr.	33.00
Filming labor	rotary camera, no automatic feed 38.5 hours @ $5.50	211.75
Inspection labor	5.1 hours @ $5.50/hr	28.05
Estimated direct costs		$339.43

To provide for the various indirect costs involved in the microfilm conversion (i.e., such one-time capital expenditures as microfilm cameras) that should properly be distributed on an amortization basis across the life of the equipment, as well as to allow for the costs of planning and supervising the micrographics application, the application of standards costing method provides for the increase of identifiable direct costs by 25% in order to provide an allowance for less tangible, less quantifiable *indirect costs*. In our example, therefore, the total estimated costs of the proposed micrographics conversion would be:

Direct costs	$ 339.43
Indirect costs (@ 25% of Indirect)	84.86
Estimated total costs	$ 424.29

Naturally, if any additional processing or formatting will be performed in connection with the micrographics conversion, the standard for these too will be developed and included in the direct costs. If we were, for example, to insert the processed microfilms into microfilm jackets, we would have to provide for the costs of purchasing the jackets, and the labor involved in their labeling and loading. Such standards may be developed based upon current costs or, in the case of hardware, the manufacturer's rated production reduced by 16.67% to provide for operator fatigue and work breaks.

Comparison with Commercial Prices

It is always sound business practice to periodically obtain estimates from commercial micrographics service bureaus for conversion utilizing their equipment and personnel. This will often indicate opportunities for savings as well as afford the micrographics manager an appraisal of how cost efficient the in-plant operation actually is. One word of caution—be certain to increase the commercial micrographics service bureau's estimate by the dollar value of any preparation, inspection, or other conversion-related tasks performed by the using organization's personnel. This will give a truer comparison of costs.

IMPLEMENTATION AND FOLLOW-UP PHASE

The initial step in the implementation of a new micrographics system is the actual installation of the new equipment and the phasing in of the micrographics-based system.

It is essential that all micrographic cameras be tested, calibrated, and adjusted as required, immediately after delivery and prior to the actual micrographics conversion. The writer's personal experience is that the various users will tend to expect much from the new micrographics system, and therefore any improperly operating equipment that might result in either undue implementation delays or in the production of less-than-satisfactory quality microimages or hard copy reproductions will have a profound, detrimental effect upon the confidence of the various users, and often will color their whole outlook concerning the value of micrographics to their own operations.

Prior to the actual phasing-in of the new micrographics system, there should be provided a period of training, debugging, and implementation during which the microforms will be created and used as a secondary reference source, with the original hard copy records continuing to serve as the primary source. This parallel operation, a common procedure in the implementation of computer-based systems, affords an opportunity to fine-tune the new system based upon actual operating conditions and user reactions.

A common mistake many users make is to not closely monitor newly implemented micrographics systems following their installation. As a result, many of the improvements that become evident after the system is in operation for several months will never be made, and the overall cost and operational efficiency will suffer. The premature abandonment of newly implemented micrographics systems is perhaps one of the main reasons why source document micrographics systems fail to reach their full potential.

It is recommended that every new micrographics system be assigned to a designated individual for purposes of monitoring its operation and to assure that the system is modified or upgraded, as required, in order to adapt the system to changing reference and use requirements and to take advantage of such technological improvements as new indexing concepts and faster, more reliable cameras.

In closing, there are three key actions that every organization should assure occurs during the micrographics system design effort:

- Be certain that the micrographics system under development and that which ultimately is recommended for implementation will satisfy the organization's unique needs and goals. There is a tendency for many organizations to try to utilize some other organization's system, on the assumption that if it works for an organization in the same or a similar business it will work for them. All too often this is not the case and the system fails to achieve its objectives.

- Many source document micrographics systems fail because too much is expected of them. This generally happens if a set of well-defined, realistic goals are not agreed to by the user and management at the very beginning of the systems analysis. The same effect occurs if the person who develops the system promises too much as the result of its implementation or in any other way oversells the system.

- After spending time, effort, and money in developing a new micrographics-based system, many organizations fail to spend just a little more to thoroughly test the new system prior to its installation and implementation. As the result, they often find that some unforeseen problem, such as too small a magnification ratio on readers and reader-printers, which could readily have been corrected prior to implementation, must now be either lived with or needless and often costly delays will be incurred while the problem is rectified. Most importantly, as described earlier, such unforeseen problems tend to adversely affect users' confidence in micrographics and often leads to resistance on the users' part when other micrographics-based applications are suggested.

7

Source Document
Micrographics Applications

BASIC APPLICATIONS

The state-of-the-art with respect to source document micrographics has advanced to the point where microfilm may be substituted for hard copy in countless numbers of applications in order to simplify and reduce the costs of the following:

- In-office storage and retrieval of records and information.
- Duplicating or publishing and then distributing records and other printed materials.
- Safeguarding records against damage, loss, or theft.
- Complying with the mandatory records retention requirements of the various governmental agencies, as well as satisfying the organization's own audit, operational, statistical, and historical reference requirements.

In this chapter, we describe the various ways in which businesses, governmental agencies, educational institutions, and nonprofit organizations of every size and type are using source document microfilm as a substitute for paper (hard copy) records, and are reducing operating costs and improving overall efficiency as the result. These applications are presented to suggest ways that the readers may integrate source eocument microfilm into their organization's recordkeeping systems *if* a systems analysis conducted in the manner described in Chapter 5 indicates that source document microfilm is the most economically and operationally efficient alternative.

IN-OFFICE STORAGE AND RETRIEVAL APPLICATIONS

Microfilm's advantage over hard copy, floppy disk, magnetic tape, and other active recordkeeping media, is its ability to maintain large volumes of records in a small filing area in what is basically a tamper-proof, human-readable format that may be readily referenced and retrieved using relatively inexpensive equipment.

Recognizing this, increasingly more organizations are looking to source document microfilm as a means of improving the cost-effectiveness of their in-office recordkeeping organization. As space and labor costs continue to rise, it is reasonable to assume that source document microfilm will assume an even more important position as an active recordkeeping media.

Microfilm's role in active recordkeeping systems is virtually only bounded by the systems designer's imagination. Source document microfilm has been used not only to increase the volume of records that may be stored in a limited space, but also to improve document accessibility, to simplify clerical processing, and to facilitate the updating of records, among other applications.

Simplification of Accounts Receivable and Customer Billing

Microfilm has enabled one medium-sized midwestern retailer, with approximately 30,000 accounts, to revise their accounting practices to utilize what they refer to as "bookless bookkeeping." Several years ago, this retailer made several significant improvements in the Accounts Receivable and Customer Billing System:

- Placed it on computer, using a local service bureau.
- Instituted "cylce billing" (so called because the preparation and mailing of customer statements are spread over the month in "cycles" for each group of customers, avoiding the end-of-the-month peaks that occur when all statements are mailed within the same week).
- Installed a total micrographics-based Accounts Receivable and Customer Billing recordkeeping system.

This system works as follows. As any of the 30,000-odd customers makes a charge purchase, pays all or part of the account balance, cancels an order, or receives an adjustment due to such factors as overpayment or return of merchandise, a suitable form, called a posting media, is prepared. At the end of every business day, all posting media are delivered to the Service Bureau, where they become the source for updating the Accounts Receivable master file (a disk file that contains the current-month-to-date status of each customer's account, and which serves as the source for producing the monthly customer statements). Additionally, this master file is used in an on-line mode, in conjunction with the

Service Bureau's COM recorder, to produce a daily transaction register, in a microfiche format. This register itemizes each charge, credit, or adjustment that was posted to the Accounts Receivable master file that particular day, arranged in account number sequence. This microfiche and the original posting media are returned to the retailer's Accounts Receivable Department. Here the posting media are interfiled with others received during the current billing cycle and are maintained in account number sequence. The microfiche is retained for subsequent reference pending issue of the monthly statements.

At the end of the billing cycle, the posting media are microfilmed using a rotary microfilm camera, and indexed utilizing code-line indexing. After processing and inspection to assure the film's quality is acceptable, the microfilm is mounted into cartridges. The original posting media are forwarded to the mail room awaiting the receipt of the monthly customer statements. The Service Bureau, at the close of the billing cycle, runs its Accounts Receivable master file and prepares both hard copy monthly customer statements and COM-generated microfiche containing copies of all such statements in that particular billing cycle. These are forwarded to the retailer's mail room.

At the mail room, personnel match the monthly customer statements with the corresponding posting media, insert them into envelopes, and mail them to the appropriate customers. The microfiche containing copies of the monthly customer statements are forwarded to Accounts Receivable, who now have only microfilm copies of the monthly customer statements and of the posting media; the individual customers have the hard copy.

As inquiries arise, the microfiche of the monthly customer statement and the cartridges containing the code-line indexed posting media provide all the information normally required to respond to the customer.

The retailer reports that the benefits from this innovation have been "significant." Customer inquiries have decreased by more than 60%, since the customer now can verify the entries on his monthly customer statement by referring to the *actual* posting media signed by him; previously, he would ask the retailer for proof of the transaction. Additionally, the microfilm-based recordkeeping system has cut the space Accounts Receivable uses for storing records needed to respond to customer inquiries by more than 90% over the old, hard copy-based system. Also, since the microfilm records may be readily and economically duplicated, each Accounts Receivable clerk servicing the same billing cycle is provided with a complete set of records, eliminating the time and money lost under the old system while traveling to retrieve the records from the files or awaiting their return by another borrower.

Increasing the Speed and Accuracy of External Audits

Through the introduction of source document micrographics, a public accounting firm has been able to:

- reduce the time required to complete their audits;
- eliminate many follow-up telephone and personal visits to the client's premises.

Utilizing a portable rotary microfilm camera, equipped with a duplex capability, the auditors visit their client's offices and microfilm those records that they require for their examination. They then return to their offices. After processing, the flash-target indexed microfilm is retained in continuous 100 foot lengths, mounted in open reels. Using inexpensive roll microfilm the auditors refer to the microimages as their examination progresses, a procedure that eliminates the "greater majority of return trips we'd ordinarily make to the client's offices to check out mistakes or re-examine some record or other."

The representative of this public accounting firm described two additional benefits that have resulted from their source document microfilm application:

- There is no possibility that they will make a mistake while transcribing information from the client's records, since the microimages are exact duplicates of the original records.
- The client will not be required to provide office space for the auditors for as long a period, since less time must be spent on the client's premises since the introduction of microfilm.

Eliminating Transcription of Information by Microfilming

Often, source document microfilm can be used for eliminating the transcription of information by photographing a single document in two separate ways. A small regional railroad's freight transfer procedure provides an excellent illustration of how this is possible.

Various freight trains would arrive every day at this railroad's freight transfer yards, where they would be broken up and dispatched to other points among a number of outgoing trains. Each freight car would be represented by one or more way bills. Under the system that was in effect before the introduction of source document microfilm, the conductor of the incoming train delivered his way bills to the Freight Transfer clerks who would then hand transcribe much of the information these forms contained into the Freight Transfer yard's "inbound register." After the various freight cars had been redistributed among a number of outgoing trains, essentially the same information would again be transcribed into the "outbound register." This manual transcription required a high degree of clerical accuracy, and the correction of errors frequently delayed the make-up and disptach of the outgoing trains. The actual way bills themselves would be delivered to the conductors of the new trains to which the freight cars were assigned.

Under the new source document microfilm procedure, the way bills are microfilmed immediately after the conductor of the incoming train delivers them to the Freight Yard clerks. They are then sorted by outgoing train and microfilmed again. The two microfilms replace the inbound and outbound registers, eliminating the time, cost, and possibility of errors involved in the manual transcription procedure. As in the original procedure, the original way bills are delivered to the conductors of the applicable outgoing trains.

Facilitating the Storage and Retrieval of Oversized, Bulky, and Nonstandard Size Documents

Considerable operating and cost problems are posed by the necessity to store and retrieve oversized, bulky, and nonstandard sized and shaped documents, such as computer print-out, maps, instrumentation graphs, newspapers, and such bound books as accounting ledgers and manuals. These records are either too large to fit into standard drawer, shelf, or lateral files without folding or are of such irregular sizes and shapes that they cannot be readily found when filed among other records that are of standard letter or legal size. As the result, the storage of such records requires either the purchase and use of such special-purpose filing equipment as tray files, stick files, and tub files, or else the folding and refolding after use of the documents to a size that can be accommodated by the filing equipment.

Microfilm can solve these problems very nicely by converting oversized, bulky, and nonstandard sized and shaped documents to a uniform sized and shaped microform. This standardization will eliminate the need for special-purpose filing equipment, make it no longer necessary that the documents be unfolded and refolded every time they are referenced and refiled, and totally preclude the possibility that a small record will be "lost" when filed among other, larger-sized records.

A county Department of Social Services located in New York State faced the problem of how to copy with the filing and retrieval problems posed by its welfare case files. These files, one of which was prepared and maintained for each recipient of any type of welfare assistance, contained an average of 37 individual records that ranged in size from $3^{2}/_{3} \times 2\frac{3}{4}$ inches to computer print-outs measuring $14\frac{7}{8} \times 11$ inches. They found that when these various records were clip-filed by means of a prong fastener into the file folder, the smaller records were positioned so far from the edge of the stack of records that a person thumbing through the file looking for a particular record would not see, touch, or be aware of the small sized records. Likewise, if the records were loosely filed (drop-filed) in the file folder, the only way to be sure that the smaller records could be located would be to remove the entire file folder and individually turn and examine each document—an extremely time-consuming, costly, and inefficient procedure.

By reducing the entire welfare case file folder to 16 mm microfilm inserted into microfilm jackets, the Department of Social Services found it possible not only to recover file space and equipment, but also to standardize the size and shapes of the various records and, most importantly, to preclude the possibility that any individual clip-filed or loosely filed document would be lost when filed among other larger documents.

Microfilm-Based Engineering Drawing System Leads to "Paste-Pot Drafting" Practices

Within the past two decades, with the introduction of the microfilm aperture card-based engineering drawing systems, drafting practices have undergone major change.

Traditionally, engineering drawings were prepared on either vellum or linen cloth using india ink, and were stored either flat in large drawing files or suspended from stick files. When duplicate copies were required for either distribution or security purposes, the applicable engineering drawing was removed from the files and was forwarded to the Reproduction Department, where a full-size, paper-based copy was prepared by either ozalid, diazo, or some other similar process. The original engineering drawing was then returned to the files and the duplicate copy was folded (a particularly time-consuming, laborious task) and forwarded to the designated recipients or the vital records depository, as applicable.

Revisions to existing drawings were processed in the following manner. The Engineering Department would initiate a change notice specifying the revisions that were to be made to a particular engineering drawing. The change notice was then forwarded to the Drafting Department, who would withdraw the original engineering drawing from the files. Using a razor blade, the draftsman would then scratch off the obsolete portions of the drawing, and revise it in accordance with instructions contained in the change notice.

After all changes had been made, the original engineering drawing would be refiled in either the flat drawing files or in the stick files. The change notice would be filed, generally in the project file, for future reference. It should be noted that with this system, unless a duplicate diazo, ozalid, or other similar copy was retained, there is no way that one may retrieve a superseded drawing other than to work back from the latest issue using each change notice in reverse order of issue, as the source.

The introduction of microfilm aperture card-based drawing systems using 35-mm microfilm led to major innovations, cost efficiencies, and simplifications in these traditional drafting practices—innovations that are commonly referred to as "paste-pot drafting techniques."

The "paste-pot drafting" concept, as practiced by a well-known electronics manufacturer, operates in the following manner. New engineering drawings are prepared on bright, opaque white bond paper, using black india ink. The drawing is forwarded to the Micrographics Department, where it is microfilmed using a planetary camera equipped with a two-inch pull down. After processing, inspection, and testing of the microimage to assure satisfactory density and resolution has been achieved, the original hard copy drawing is refiled in the flat drawing files. The corresponding microimage is cut to size and mounted in a microfilm aperture card that has been keypunched and interpreted to include such indexing data as the drawing number and title, the revision number and date, and the applicable project number. Two duplicate, similarly keypunched and interpreted diazo copies are prepared from the original microfilm aperture card. One copy is filed in the "slave deck" for subsequent mechanical sorting and retrieval. The other copy is transmitted to the vital records facility, where it is maintained for retrieval should the original microfilm aperture card be lost, misfiled or be otherwise unavailable. After preparation of the diazo duplicates, the original microfilm aperture card is filed in drawing number order in the microfilm engineering drawings files.

When it is necessary to retrieve a particular drawing, the original microfilm aperture card is removed from the microfilm engineering drawings files, matched with a blank diazo duplicate aperture card, and passed through a card-to-card duplicator. The original is then refiled; the diazo duplicate is forwarded to the requestor, who views it in a suitable reader or reader-printer, making hard copy reproductions on the latter hardware should they be required.

Revisions to existing engineering drawings are initiated by the Engineering Department, who prepare a change notice and forward this form to the Drafting Department as their authority to revise the drawing. A draftsman removes the hard copy drawing from the flat drawing files, and pastes a piece of bright, opaque, white bond paper over that portion of the drawing that is to be revised. Using a pen and black india ink, he then revises the drawing. The change notice is filed in the project file; the revised engineering drawing is forwarded to the Micrographics Department where it is microfilmed on a planetary camera equipped with a two-inch pull down attachment. The microfilm is then processed, inspected, and tested. If acceptable, it is cut to individual microimage size and mounted in a microfilm aperture card that has been keypunched and interpreted as before. Two diazo duplicate copies are then prepared; one is forwarded to the vital records facility, the other included in the "slave deck." The original microfilm aperture card is then filed in the microfilm engineering drawing files; the superseded original being removed and refiled in the historical drawings file in drawing number and revision data order.

The installation of this microfilm aperture card-based system has enabled the electronics manufacturer to reduce the personnel and materials costs related to the revision of engineering drawings by over 75%. Additionally, the following benefits were realized: the costs for filing equipment, supplies, and floor space to house such filing equipment decreased by over 20%; the ease and low-cost of maintaining an historical file of superseded engineering drawings enabled the manufacturer to quickly and accurately reconstruct and retrieve such drawings.

Facilitating File Integrity

Records are the memory of any organization. They document actions taken in the past and those planned for the future. Yet, to be useful, the information contained in records must be current, complete, and retrievable on demand.

Hard copy-based files have historically been plagued by four file integrity problems:

- Misfiled records.
- Borrowers failing to return documents.
- Alteration and/or obliteration of information contained in the records.
- Borrowers failing to properly complete an ''OUT'' card whenever they remove a record, group of records, or an entire folder from the file cabinet.

As a result, confusion, a loss of time, and needless expense often ensues. Retrieval requests may be delayed or may even go unanswered as file clerks search for missing or misfiled records that contain the desired information.

A Connecticut-based life insurance company's Investment Department maintains, as one of its primary reference tools, a file on each of the 6,000-odd corporations whose stock is traded on either the New York or the American Stock Exchange. These files, called ''company files,'' contain a plethora of information relating to each listed company's financial position, operations, and management, as well as copies of any analyses of the company's current and future investment potential that may have been prepared and issued by various investment advisors, brokerage firms, or banks.

These company files were the primary source for researching potential investments and were used by all the Investment Analysts in the department. They therefore had to be kept up-to-date and complete. In a hard copy-format, such file integrity was extremely difficult to maintain. Analysts would often forget or neglect to sign out a file; or would fail to return the entire file, keeping one or more records for ''follow-up'' and not informing the file clerks that they had done so. As the result, users could never be certain that the file was either current or complete, and occasionally made decisions based upon partial or noncurrent information.

Ultimately, it was decided to convert the company files to a microfilm jacket format. Each Investment Analyst was provided with diazo copies of whichever company file they required, which they could then reference on low-cost microfiche readers. Once their immediate reference needs were satisfied, borrowers were instructed to discard the diazo copies, precluding the possibility that an obsolete, noncurrent record would find its way into use. The low cost of the diazo copy (approximately 20 cents each) made this a viable alternative.

The microjacket system alleviated the problems of economically and conveniently providing and maintaining file integrity for this life insurance company, and has greatly increased the level of confidence the various Investment Analysts place in the completeness and currency of the company files.

Assuring Document Accessibility

The development of central files and centralized records collections results in numerous economies of scale and efficiencies that normally cannot be justified or obtained were the records maintained on a decentralized basis, for example, the maintenance of the files by qualified, full-time file clerks, rather than by secretaries and file clerks as an additional duty, the use of computer-based indexes and of mechanical filing equipment. However, centralization invariably results in accessibility problems. If a record is maintained at a centralized location, the borrower must either travel to that location or the requested record must physically be transported to wherever the borrower is located. There is also the possibility that the same record may be simultaneously requested by two or more individuals or that someone may request a record while it is out on loan.

A properly designed microfilm system will alleviate these problems in one of two ways. The first method, which has been implemented by a multibranched savings and loan association, is to convert the centralized records to a microfilm format (either continuous lengths of microfilm in cartridges or microfilm jackets, depending upon whether a unitized or a nonunitized microform is required). These various microforms are then duplicated in their entirety, and a satellite file of diazo duplicate continuous roll film in cartridges and microfiche is set up at whichever decentralized location's reference requirements necessitate access to such records.

The second approach, which is used by the Technical Library of a manufacturer of refractories, is to respond to requests from Field Sales Engineers for details concerning the specifications and properties of various products, which the Technical Library maintains on a centralized bases in microfilm jackets and microfiche format, by creating a duplicate microfiche by fiche-to-fiche diazo duplication. The diazo duplicate is then sent to the requestor either by messenger or mail, depending upon the distance involved and the speed of response required.

Using Microfilm to Furnish Subpeonaed Records

A chemicals manufacturer, after being named a co-conspirator in a price fixing case, received a subpeona from the Department of Justice that necessitated the retrieval and delivery (to the Department of Justice) of over 80,000 letters and memoranda relating to their sales and marketing efforts. Since most of the subpeonaed records were maintained in just one copy, it would be necessary for the manufacturer to prepare two copies of each subpeonaed document—one for his own use and one for his legal counsel's.

It would have been a monumental, extremely costly task to create and collate over 160,000 individual copies using any of the standard office copying methods, so alternative solutions were sought. After analysis, it was decided that the use of microfilm as a copying medium would prove most cost-effective. Each individual letter or memorandum was microfilmed using a rotary microfilm camera equipped with an automatic document feed feature. After processing and inspection to assure quality of the microimages, the microfilm, in continuous roll form, was passed through a high-speed printer, which produced full sized hard copy enlargements of each microimage, in speeds averaging 20 letter sized, hard copy enlargements per minute. Since two copies were required, the process was repeated creating two separate sets of the subpeonaed records. One set was retained by the manufacturer, the other delivered to his legal counsel. The original letters and memoranda were forwarded to the Department of Justice.

This utilization of micrographics not only reduced the costs of preparing copies of the subpeonaed records, but it enabled the manufacturer to meet the submission date specified in the subpeona, an accomplishment that it was felt would not have occurred were standard office copying equipment used.

Use of Microfilm to Reconstruct Water-Damaged Records

During a recent summer, the Sckuykill River in Philadelphia overflowed its banks. After the waters receded, a textile firm returned to find that its technical information files, which were located on the ground floor had been inundated with water, and all the documents contained therein were thoroughly soaked. Since much of the information in these files were of a proprietary nature, or represented projects under development, the records represented a large portion of the firm's "know-how" and could not be duplicated from other sources. The firm had two alternatives: either undertake the slow, costly, space-consuming task of drying out the water-soaked records, or prepare duplicates of the water-logged records, retaining such duplicates, in lieu of the damaged originals, for future reference. Clearly, time and cost factors ruled in favor of the latter alternative. After investigation, it was found that the use of a planetary microfilm camera to create microfilm copies of such records would prove the most economical, satisfactory procedure. Records were lifted up carefully, assisted by a standard cooking spatula, and placed on the flat bed of a planetary camera. After the necessary lighting adjustments, the record was microfilmed and placed in a carton until the microfilm had been processed and found to be of satisfactory quality. The original, water-damaged records were then destroyed, and the microimages retained for future reference. This procedure reduced the time and costs attendant to the reconstruction of the water-damaged records by over 60%.

Reducing the Costs of Relocating Records

The costs involved in the moving of records from one building or city to another, such as would occur during an office relocation, will typically range upwards from $1.50 per file drawer of records for local moves to $5.00 or more per file drawer for long distance relocations. Additionally, labor costs will be incurred in preparing the records for their relocation and setting them up in their new location.

A pharmaceutical manufacturer in the process of relocating their administrative offices from a large Northeastern city to a suburban location some 50 miles southwest, found that significant costs savings would be realized were they to reduce records to microfilm before relocating them to their new location. Specifically, the cost of transporting the records alone was $2.00 per file drawer, and by means of a sampling it was found that approximately 30 minutes labor would be involved in preparing a single file cabinet of records for transfer and setting it up at the new suburban location. The manufacturer's cost analyses showed that were the records microfilmed prior to their relocation and retained until their normal destruction dates in microfilm rather than a hard copy format, the costs of moving and retaining the records would be reduced by nearly 45% as the result of:

- reducing the bulk of the records to be moved, and therefore their moving costs;
- eliminating the possibility that the records will become disarranged in transit and a time-consuming, costly rearrangement will therefore become necessary;
- reducing the floor space and filing equipment required at the new location for the storage of the records.

Maintenance of Records Required at Irregular Intervals

The maintenance of records such as medical case files, project development files, and policyholder claim files, in which it is virtually impossible to determine when the records will be needed for immediate reference and which by reason of their volume cannot be economically maintained in costly office areas, is another excellent application for source document micrographics. Such files can be reduced to microfilm and, depending upon their volume and the future possibility of updating, be formatted in cartridges, microfilm jackets, or as microfiche, to be retained in office space pending future use. Were such files maintained as hard copy, it would be impossible to store them indefinitely in readily accessible office files; they would of necessity have to be transferred to a low-cost inactive records storage area.

Such an arrangement could well lead to operational difficulties. In the case of medical case files, this becomes readily apparent, for in the event of admission to a hospital at which he had previously been a patient, the applicable medical case file must be retrieved and reviewed in order to better plan for treatment. Microfilm is a natural in such a situation, since it allows large volumes of records to be maintained for long periods in limited office space. The system in which microfilm is used is as follows:

- When a patient is admitted to a hospital, and during the course of his stay, numerous forms, charts, and other records are created detailing his malady and its treatment. These remain at the patient's bedside during his hospitalization.
- Upon discharge, the various medical forms, charts, and other records are forwarded to the Medical Records Department where they are microfilmed.
- After inspection to assure satisfactory quality, the microfilm is inserted into a microfilm jacket that has been labeled with the patient's name, address, and other identifiers. This microfilm jacket is then filed in the Medical Records Department's medical case files.
- The original hard copy medical forms, charts, and other records are then destroyed.
- Should the patient be readmitted, his microfilm jacket-based medical case file is retrieved. A diazo or vesicular duplicate is prepared and is forwarded to the appropriate physicians for reference and use in planning for the treatment of the current malady or condition.
- Again, various hard copy medical forms, graphs, and other records will be created and retained as hard copy during the patient's confinement. Upon his discharge, these hard copy records will be forwarded to the Medical Records Department where they will be microfilmed and added to the patient's microfilm jacket-based medical history file.

8

Computer-Output Microfilm

Computer-output-microfilm (commonly known as COM), is a process that enables digitized data to be converted from a computer-processible format to either continuous lengths of microfilm mounted on reels, or in cartridges or cassettes, or to microfiche, without first being printed out as hard copy. As such, COM is an electronic data processing (EDP) output technique that provides an alternative to either hard copy print-outs that are produced by either impact line printers or the newer, faster nonimpact printers, or to the CRT-based visual display of data.

COM AS AN ALTERNATIVE TO OUTPUT PRINTERS

COM offers an attractive, cost-reducing alternative to either impact line printers or the newer, faster nonimpact printers in those applications in which the production of reference and working copies of EDP output forms and reports are required:

- COM eliminates the purchase and use of customized hard copy forms as well as of stock tabulating forms.
- With COM, such time-consuming, costly forms handling operations as decollating, bursting, and binding are not required.
- The use of COM results in materials savings of approximately 90% over hard copy.
- COM's output rate is approximately 20 times faster than the average EDP

impact line printer's and about 10 times faster than the typical nonimpact printer's. This leads to a reduction of up to 90% in computer operating time.

- Records maintained in a COM format may be duplicated at a cost that is nearly 30% lower than the cost of duplicating hard copy output produced by EDP output printers, of either the impact or nonimpact variety.
- Due to its compactness, COM reduces the space required to store computer-generated output by 98% if reels, cartridges, or cassettes are used, and 99.5% if microfiche is involved.
- The costs of distributing records in a COM format via either the U.S. Postal Service or through internal mail, will be approximately 90 to 95% lower than would be incurred were those same records maintained and distributed as hard copy.
- Down time has been found to be significantly less for COM devices than for EDP output printers.

The net effect of the foregoing efficiencies and economies, according to a recent study conducted by U.S. Datacorp, is a reduction in "the cost of computer output [of] anywhere from 35 to 50%, depending upon the application."

Lest one conclude that COM is superior in all ways to hard copy EDP output, there are a number of limitations and disadvantages that must be weighed against these preceding advantages:

- It is not possible to utilize COM as the sole method of preparing EDP output, regardless of the potential cost and operational efficiencies. Many computer-generated forms and reports, such as sales invoices, must be forwarded to outside organizations and customers. Since it is unlikely that these recipients would have hardware capable of reading COM, it will be necessary to continue to produce hard copy forms and reports. In many applications, therefore, the system will be a hybrid one in which COM is used for internal distribution purposes, and output printer-generated hard copy is distributed to outside organizations and to customers.
- COM does not lend itself to browsability as readily as hard copy does.
- Neither can COM be annotated, as can hard copy.
- Since reference to COM requires a suitable reader, it is not as portable a recordkeeping media as is hard copy.

COM AS AN ALTERNATIVE TO CRT's

For those applications in which a data base must be stored, updated, and accessed, the choice of output media is basically between CRT-based visual display and COM. COM offers a number of significant advantages over the latter output alternative:

- Its use requires only minimal programming. CRT-based visual display, on the other hand, necessitates the development of special-purpose, often complex software.
- There is no limit to the number of users who can be furnished their own retrieval equipment when COM is involved. With CRT-based visual display, however, there is a maximum number of CRT's that may be linked to a single computer. In all fairness, however, this limitation may prove merely academic, since as many as 200 CRT's may be linked to a single computer.
- COM requires no special wiring or communications linkage: CRT-based visual display units must be connected to the computer by means of either cables or over telephone lines.
- COM readers cost considerably less to purchase or lease than do CRT-based visual display units.
- COM readers are much more portable than are CRT-based visual display units.

These alternatives must, however, be weighed against the following disadvantages:

- Users have the ability to directly interact with the computer, and update the data base themselves by means of the CRT-based visual display units. This feature, which assures timeliness of information at all times, is not a comparable feature of COM.
- Response time is faster with CRT-based visual display than with COM.
- There is greater security with CRT-based visual display. The applicable software may require that users furnish specific access or authorization codes before all or certain specified data contained in the data base will be made available to them and be displayed on the CRT. With COM, on the other hand, access to the microforms and a suitable reader affords one access to all the information contained in that microform.
- There is no waiting period between the time the data base is updated and the availability of that information to the various users when CRT-based visual display is involved. With COM, there is such a delay while the microforms are prepared, duplicated, and distributed.

The selection of either CRT-based visual display or COM, therefore, generally involves a trade-off between the costs and references and use requirements. As a general rule, if the application necessitates continual updating and the immediate availability of current information to all users (what systems analysts refer to as "real-time applications"), CRT-based visual display is the only viable alternative. If, however, it is possible for users to utilize information that reflects the data base as it existed at some specific previous point in time (such as at the close of business the previous day), COM will prove a satisfactory alternative.

TYPES OF COM DEVICES

There are two primary types of COM devices:

- *Alpha-numeric printers,* the type most commonly used in business applications. Such COM devices produce alpha-numeric characters and a variety of symbols (such as the ampersand, comma, period, and asterisk, as well as such mathematical symbols as $+$, $-$, and $\times$, among others) on microfilm in a standard computer output format of 132 print positions per line and 64 lines per page.
- *Alpha-numeric printers/graphic plotters,* which in addition to the alpha-numeric capability described above, can also produce such graphics as bar charts, graphs, engineering drawings, and scientific plots.

BASIC COMPONENTS AND OPERATION OF COM DEVICES

Although there are considerable differences among the various COM devices in use today, insofar as the type processor utilized, the format of the output, the method of character generation, and the like, it is possible to describe generically the method by which all COM devices convert computer-processible data to microforms. The COM device itself consists of six basic components: The input section, the logic section, the conversion section, the deflection controls section, the display section, and the film handling section. The actual operation of the typical COM device is as follows:

The *input section* receives digitized data directly from either the computer's central processing unit (CPU) in an on-line operating mode or from an auxiliary storage device such as a tape drive in an off-line mode. The digitized data is converted to an electrical signal and is transferred to the *logic section,* where interpreting and formatting occurs. The signal then passes to the *conversion section,* where they are converted to analog signals and then to alpha-numeric characters, lines, symbols, and the like, as applicable. These characters, and so forth, are next transferred to the *deflection controls section,* which adjusts their positioning on the face of the COM device's visual display, or in the case of some devices on the face of the microfilm itself. The *display section* receives and holds the image for filming, while the actual microfilming is performed by the *film handling section,* which is composed of the lenses, film transport mechanisms, and other items required for actual microfilming.

A unique feature of the film handling section is the ability to insert a blank forms overlay between the camera's lens and the display section. Since this overlay contains the lines, captions, and other fixed data contained in the computer-generated form or report, and is fully transparent, when microfilmed it creates a microimage that includes all fixed and variable data that would be included in the hard copy report or form.

COM RECORDING TECHNIQUES

There are four techniques by which the various COM devices may receive and convert digitized data into ordinary alpha-numerics, lines, and symbols, and then produce such converted data on microforms:

- *CRT Recording,* in which the alpha-numerics, lines, and symbols are projected on the face of a cathode ray tube included in the display section for subsequent microfilming.
- *Electron Beam Recording,* a technique in which the various alpha-numeric characters, lines, and symbols are "written" directly onto the face of dry silver microfilm.
- *Light Emitting Diode Recording,* a technique that directs the signals to a bank of light emitting diodes (LED's) arranged in a matrix. The light from the LED's is passed through and is concentrated by bundles of fibre optics to form an alpha-numeric character, a line, or a symbol, as appropriate, which is in turn photographed by the film handling section.
- *Laser Beam Recording,* a technique that uses a laser beam to "write" the applicable characters, lines, and symbols directly onto the face of a dry silver or vesicular microform.

COM FORMATS

Depending upon the particular COM device employed, COM output may be formatted as either 16- or 35-mm wide continuous lengths of microfilm mounted on open reels, or in cartridges, or as 148 × 105 mm microfiche. Individual microimages (or frames) may be arranged in either a comic or cine mode, at reduction ratios of 24×, 42×, or 48×. Figure 8-1 summarizes each of these formats.

The specific reference and use requirements of the individual application is the primary determinant as to which COM format will be selected. As a general rule, the selection of a microform must be based upon an analysis of the following factors:

- The number of microimages (or frames) involved.
- The need to subdivide these total microimages to satisfy the reference and use requirements of individual users of the COM-generated microforms.

As a general rule, a large, continuously referenced COM-based file will be

COM FORMATS

Format	Reduction Ratio	Maximum Number of Micro-Images of 8½"x11" Originals
16mm Film	24x	2,400 Per 100 Feet
	42x	4,200 Per 100 Feet
35mm Film In Roll Form	24x	4,800 Per 100 Feet
	42x	16,000 Per 100 Feet
148mm. x 105mm Microfiche	24x	98 Per Microfiche
	42x	325 Per Microfiche
	48x	420 Per Microfiche

Figure 8-1   COM formats.

referenced faster and easier if maintained in an open reel or cartridge format than if produced as microfiche. This becomes evident when one considers that the more than 4,000 microimages (frames) that can be maintained on a 100 length of microfilm would require over 20 individual microfiche, if that latter microform was utilized. If the user need only insert a single reel or cartridge of microfilm in a reader before conducting a series of references, the look-ups will be completed faster and at less cost than if individual microfiche had to be inserted into and removed from a reader each time a reference was conducted.

If, however, the COM-based file must be subdivided in order to provide various users with specific records (e.g., all records relating to customers whose account numbers fell within a given numerical range) or if the volume of the

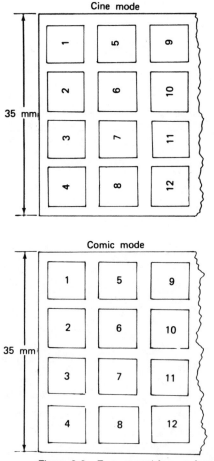

Figure 8-2   Two-up and four-up formats.

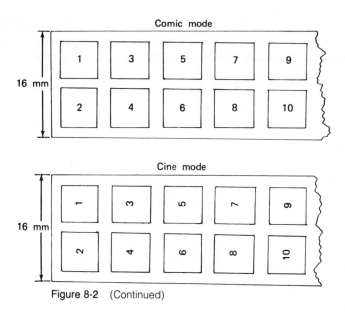

Figure 8-2   (Continued)

COM-based file is less than 1,000 microimages (frames), then microfiche will generally prove the faster, easier format to work with.

It is possible to increase the storage capacity, or "density" as it is more properly called, of 16- and 35-mm microforms by programming the COM device to arrange the various microimages in a two-up format at reductions of 42× or more across the width of 16-mm microfilm or from two-up to four-up across the width of 35-mm microfilm (Figure 8-2).

COM-generated master microforms are generally of a positive polarity, producing an image that contains black images appearing upon a white background. Even though silver halide negative microfilm is used by most COM devices, this phenomenon is possible as the result the photography methods used by COM devices. Using a COM device, the only light is that which falls upon the face of the CRT and is photographed by the COM device or which is written directly on the face of the microfilm by the COM device's electronic beam or laser beam. There is no reflected light coming off the background, as is the case in standard source document microphotography. As a result, after standard chemical processing, a positive image is produced.

Generally, duplicate distribution copies of COM-generated microimages are produced using diazo film, which, as described, produces an image of the same polarity as the original microform.

However, many users prefer working with a negative microform, finding that less glare results. To obtain such negative microforms, duplicate copies are generally prepared using a vesicular film, which as described in Chapter 4 yields an image that is opposite in polarity to the original, or else produces a negative silver halide master from the original positive one by means of a process known as "full reversal processing" and references either that duplicate or uses it to produce additional diazo duplicate microforms, each of which will be of negative polarity.

LEGALITY OF COM

Many state and governmental agencies have provided for the maintenance of information required for their examination in a COM rather than a hard copy format should the user so desire to exercise this option. As a rule, such regulations specify minimum indexing and quality standards, and mandate that a suitable reader be made available for the agency's personnel to use in reviewing the COM-based files. It is wise to check with the various agencies to whose jurisdiction your own organization is subject before converting any records required by that agency from hard copy to COM-generated microforms to assure that there are no prohibitions against your doing so, as well as to assure that you have taken such requirements into account in your system design.

Some organizations have included a Certificate of Authenticity in their COM-generated microforms in the belief that it will facilitate the admissability of those records as evidence in courts of law and in regulatory proceedings. Figure 8-3 is an example of one such Certificate of Authenticity, which appears on the envelope in which the COM-generated microfiche are filed.

SELECTING A COM DEVICE

The selection of an in-house COM device necessitates an evaluation of the user's unique informational and processing requirements. Only then will it be possible to select the COM device whose characteristics most closely approximate these requirements.

The selection process is one of identification and elimination, as first the requirement is defined and then the particular COM devices that can satisfy that requirement are identified. Little by little, the choice of available COM devices will be narrowed until only a few are left. At that point, a judgment must be made as to which will most nearly approximate the user's requirements, at the lowest cost. Undoubtedly, there will be a trade-off, since rarely will one COM device prove capable of most efficiently and economically satisfying any user's requirements.

TENNECO

MICROFICHE
DO NOT BEND

TAPE NAME		TAPE NO.(S)	
SNUMB	☐ MASTER ☐ COPY	OPERATOR NO.	ENVELOPE SEQ. NO. _____ OF _____

CERTIFICATE OF COMPLETENESS AND AUTHENTICITY

This is to certify that the _____ enclosed microfiche pieces containing Computer Run/Report
(Quantity)
entitled _____

dated _____ for _____
were produced in the regular course of business pursuant to authorized Company practice of recording this
computer produced record on microfiche and storing these microfiche in protected locations.

It is further certified that these microfiche are accurate and complete reproductions of computer data
which would otherwise have been printed as a full size report.

It is further certified that the microphotographic processes were accomplished in a manner and on film
which meets with requirements of ANSI Specifications for permanent microphotographic output.

COMPUTER CENTER		CORPORATE RECORDS MANAGER	
SHIFT SUPERVISOR	DATE	AUTHORIZED SIGNATURE	DATE

TEN 4876 3/76

Figure 8-3 Certificate of Authenticity, COM. Courtesy Tenneco, Inc.

The first task is to define the type of COM device that is required. This can be most expeditiously done by defining the type of output that will be required. If the COM device will produce only alpha-numeric reports, forms, and other records in a microformat, then one may narrow further consideration to alpha-numeric printer-type COM devices. On the other hand, if the output is to be graphic as well as alpha-numeric, one of necessity will have to consider the more costly alpha-numeric printer/graphic plotters-type COM devices.

The next selection criteria is to identify the format of output required in terms of:

• microform,
• image orientation,
• reduction ratios.

As described before, each COM device has its own unique combination of formats that it can produce. Some produce only 16-mm continuous lengths of microfilm; others only 105-mm in a microfiche format. Some are capable of producing microimages in both roll and microfiche, in a variety of reduction ratios. Some yield a comic mode, other a cine mode, still others can produce microimages in either orientation. By defining each of these output factors, a further refinement of available COM devices will result.

Next, the indexing requirements should be identified. This too will likely lead to the identification of certain COM devices that cannot provide that type of indexing.

The determination should then be made as to whether the COM device will be operated in either an on-line or an off-line mode, and the equipment that provides for this type of operation identified.

By this time, the potential COM devices will be greatly reduced in number, and the selection procedure can procede to evaluate:

- the throughput speeds of each unit,
- the type of processing required for the master microform,
- the operating cost,
- the software that is required to support the system,
- recording technique employed,
- available options, such as forms overlays.

After narrowing the potential COM devices to one or two, a trial run should be arranged in which actual operation is conducted, and COM conversion actually occurs. Live data should be input into the COM devices under evaluation and the output carefully analyzed to see how well it satisfies the users' requirements. Opinions should be solicited from a wide range of user departments to assure that as many unique requirements as possible are taken into account before one specific COM device is selected.

USE OF A COMMERCIAL COM SERVICE BUREAU

Comparatively few organizations choose to embark on COM by establishing an in-house facility. Most employ the services of a local COM Service Bureau for the conversion, processing, and duplicating of COM output. This method enables their low-cost entry into COM, and enables them to gain experience with COM systems and applications in general, before undertaking so significant a step as installing their own COM device.

Additionally, the use of a COM Service Bureau in the formulative stages of COM developments enables an organization to avoid having an expensive item of hardware sit idle while awaiting new applications. It also precludes the possibility that the state-of-the-art will lead to the development of improved equipment that will leave the organization with obsolete, less desirable COM devices.

SELECTING A COM SERVICE BUREAU

Evaluation Criteria

The selection of a COM Service Bureau requires careful, detailed study and considerable fact finding and interviewing, since the failure to select an organization that can consistently provide a high-quality product, on time, could well undermine even the most meticulously planned COM application. The author recommends that each potential COM Service Bureau be evaluated on the basis of the following five factors:

- quality of product,
- technical capability,
- reliability,
- willingness to suggest new applications,
- comparative costs of services.

Each of the COM Service Bureaus being evaluated as possible sources of COM services should be required to submit the names of six current and three former clients and of the person in each of those nine organizations who can best address the question of quality of the COM Service Bureau's product, performance, and personnel. Each of these individuals should then be interviewed by telephone in order to obtain their experience and the degree of satisfaction with the applicable COM Service Bureau. The author recommends asking each interviewee to use the following ratings:

- consistently excellent,
- satisfactory,
- less than satisfactory.

Quality of Product

In determining the quality of product produced by each of the COM Service Bureaus, interviewees should be asked to rate the following with respect to both COM-generated master microforms and all duplicate distribution and working copies made from such masters:

- clarity of the alpha-numeric characters,
- the density or contrast that exists between the background and the alpha-numeric characters,

- the readability of the titles and other identifying information that is contained in the header of the COM-generated microfiche (naturally, this factor is not applicable if a roll output is involved),
- the registration of the microimages,
- color quality,
- cutting of the COM-generated microfiche (again, this is not applicable if the output is in roll format).

Ideally, all alphabetic characters and numerals should be sharp and clear. Good contrast should exist between the background areas and all characters. The header area should be very easily read. Registration should be precise, with no noticeable skewing. There should be no signs of either discoloration or improper cutting.

Technical Capability

Technical capability refers specifically to the COM Service Bureau's equipment (hardware) and the competence and knowledge of its personnel. The various COM devices, processors, and other equipment that is utilized by the COM Service Bureau should be representative of the state-of-the-art, and its personnel should possess a level of technical expertise that will assure their ability to satisfy their clients' current and anticipated future needs.

Reliability

COM Service Bureaus, by definition, must be reliable and totally responsive to client requirements. Delivery schedules must be religiously adhered to, and all promises with respect to priority and emergency services must be met. Specifically, during the interviews with current and former clients, determine each of their turn-around requirements in order to judge if any of the problems they relate with respect to reliability may have been caused by an overly anxious salesman who may have promised an unrealistic turn-around time.

A nucleus of long-term clients is perhaps the best indicator of a COM Service Bureau's reliability. It is important, therefore, to learn how long each client has been dealing with the COM Service Bureau and (in the case of the three former clients) if poor reliability was a motivation behind their seeking a new source of COM services.

Willingness to Suggest New COM Applications

The COM Service Bureau that is truly servicing its clients will routinely suggest new COM applications, regardless of their client's size or the volume of his business. A large San Francisco department store explained that their COM Service Bureau's sales representative periodically calls upon them to see if he can be of assistance in suggesting or planning for new COM applications, and will assist in the design and testing of such applications if the client requests.

A COM Service Bureau is an excellent source of cost reducing, work simplifying ideas. Its willingness to share these ideas with its clients should, therefore, be a key evaluative criterion.

Comparative Costs of Services

Intense competition, coupled with the availability of numerous in-house COM devices, has kept the prices charged by most COM Service Bureaus, with whom the author has come into contact, within a given, highly competitive range. Rarely, in any given geographic area will COM Conversion Costs vary more than 1/10th of a cent per frame (or microimage) from one COM Service Bureau to the next. Still, it is essential to assure that each of the COM Service Bureaus being evaluated as potential vendors charges competitively for its services.

WHEN SHOULD AN IN-HOUSE COM FACILITY BE CONSIDERED?

The point at which an in-house COM facility can be cost-justified will be dependent upon the prices paid for COM conversion, duplication, and other services. As a general rule, the economic break-even point for an in-house COM facility is not reached until *COM conversion costs alone* approximate $2,600 monthly—or until the number of frames converted to microfilm via COM exceeds 205,000 each month.

9

Computer-Output-Microfilm Applications

As organizations search for new, improved methods and procedures as a means of reducing operating and overhead costs, their use of COM as an alternative to hard copy computer output is increasing. Today, in the author's experience, it is not uncommon to find that within 2 years after an organization has installed COM applications between 10 and 20% of their total hard copy computer output will have been converted to a COM format. Knowledgeable industry sources have estimated that as much as 35% of the typical medium-to-large computer user's forms, reports, lists, and other output will eventually be produced as a COM-generated microform.

CHARACTERISTICS OF COST-EFFECTIVE COM APPLICATIONS

As a general rule, COM may be economically and efficiently substituted for computer-generated hard copy whenever the following reference and use conditions are found to exist:

- The records under study are basically self-indexing; being sequentially arranged in either alphabetical or numerical order.
- A large volume of computer-generated output is produced with each issue.
- Such data is continually updated.
- At least two copies of the computer-generated forms, report, lists, and the like, are distributed within the originating organization.

- The records will not be manually annotated following their issue.
- Minimum browsability is required.

Naturally, there will be instances in which cost-effective COM applications will result if one or more of the above conditions are not present (e.g., if only one copy of the computer output is distributed within the organization), so any potential application that meets the greater majority of these conditions should be considered as a candidate for a COM application.

COM applications span the complete spectrum of business activities. Currently, industrial, financial, governmental, educational, and nonprofit organizations of every size and type are utilizing COM in their accounting, engineering, personnel, research, production, and other recordkeeping operations. To enable one to fully appreciate the versatility, scope, and potential of COM, we shall examine some of the applications in which it is now being used.

BASIC COM APPLICATIONS

There are three basic applications for COM:

- Replacement for the output printer in the preparation of computer-generated forms, reports, listings, and other textual materials.
- Preparation of such business graphics as bar charts, PERT Charts, and financial graphs.
- Production of graphic materials in which accuracy of detail and a quality microimage is essential, such as in engineering drawings, schematic diagrams, and electronic circuitry diagrams.

COM AS A DIRECT REPLACEMENT FOR OUTPUT PRINTERS

In this, the most straightforward and common of all COM applications, computer-generated forms, reports, listings, and other textual matter is produced in a microfilm format rather than as hard copy, and recipients are furnished a compact, miniaturized microform rather than a bulky hard copy. Providing that the recipient's reference and use requirements can be satisfactorily met by the microform, this application will typically result in significant cost reductions and in increased operating efficiencies. The following have been selected to provide the reader with some indication as to how COM has replaced the output printer as a means of producing computer-generated data.

Banking

Daily Transaction Registers

Daily transaction registers are produced for virtually every bank department that receives or disburses money or securities. The Installment Loan Department, for example, produces such a register to record any new loans written and payments that have been made on existing loans. Likewise, the Demand Deposit Accounting Department indicates in their daily transaction register all the debits and credits that processed on a specified date. Producing the daily transaction register is a costly, time-consuming operation, necessitating considerable computer printing and forms handling. Since these records are generally maintained for 6 months or longer, the costs of storage space and filing equipment quickly mounts up. COM has been very successfully utilized to simplify and reduce the costs involved in producing, distributing, and maintaining the various daily transaction registers. The typical COM application for such registers is as follows:

- The current day's transactions are input into and processed by the computer, but instead of being printed out as hard copy, it is transfered to disk storage.
- At the conclusion of the day's computer processing, all data relating to that department's transactions are transferred from disk storage to a COM print tape.
- The print tape is then fed, in an off-line mode, into a COM device, producing the daily transaction register.
- After processing, and formatting, the required number of distribution copies are prepared, using a suitable diazo or vesicular roll-to-roll or fiche-to-fiche duplicator.
- Each recipient is forwarded his own microform copy of the daily transaction register, which he may subsequently reference using a suitable reader or reader-printer.

Customer Statements and Receipts

Banks also use COM in both their trust and banking functions in order to simplify and reduce the costs of preparing, distributing, and maintaining such customer statements and receipts as the Personal Trust cash statement, Corporate Trust's stock transfer listings, Safekeeping's collateral receipts, and Personal Banking's monthly statement of customer account. Typically, such records require the production of both a hard copy for the customer, and one or more microfilm copies for the bank's internal distribution. Rather than prepare the hard copy by standard output printer techniques, many innovative banks utilize COM-generated microfilm and a high-speed contact printer for this purpose. By superimposing a forms overlay on their COM device's display unit, a microimage is produced that resembles a hard copy form in both design and content. After processing, the microform containing these microimages is passed through a high-speed contact printer, yielding an enlarged hard copy of the statement or receipt form. The hard copy is mailed to the customer; the original microform is used as a master to produce the number of duplicate microforms required for internal distribution.

Manufacturing

Due to the sheer volume of transactions and stockkeeping units involved, most hard copy-based inventory control systems produce stock status reports and inventory history reports once a week. Between the running of these reports, any revisions necessitated by the receipt or issue of stock items, changes in prices, or the like, generally are manually posted to the hard copy reports. With such a system, both the frequency of errors and clerical costs are high. Due to the time and expense involved in producing the reports by standard output printing techniques, however, manufacturers had little choice but to continue this admittedly inefficient system. The introduction of COM, however, has reduced the time and costs involved in the off-line production of the stock status report and inventory history report to where daily updates are both operationally and economically feasible. As a result, users are furnished more timely and accurate inventory control data.

Mail Order Marketing

Customer Service

Approximately 60% of all inquiries received by the typical mail order marketing company asks: "Where's my order?" Prior to COM, replying to such a simple, straightforward request necessitated the maintenance of voluminous shipping registers, each containing a single day's outgoing orders, as well as a second file of items on back-order. Clerks ascertaining the status of a given order would be faced with the laborious task of searching one or more day's shipping registers to determine if the order had been shipped. If it had not been, they would then reference the second source to see if it was on back-order. Only then could they confidently answer the customer's simple question. COM, with its rapid throughput and high storage density, makes it possible to produce a daily cumulative order register, in which each item ordered by every customer since the beginning of the month is individually listed, along with its current status (i.e., on back-order, shipped, or open). Now all the clerk responding to a customer's "Where's my order?" inquiry need do is look in one place for all the data required to determine any order's current status. She need only remove a small number of sequentially arranged microforms containing alphabetically or numerically arranged microimages, select the appropriate microform into a reader or reader-printer located at her work station, locate the specific frame (microimage) that contains the inquiring customer's orders and complete her reference without further delay, giving the customer an answer while she waits on the phone.

Sales Organizations

Price Lists

The maintenance and issuing of price lists is another task made complicated by the sheer number of entries involved and the computer and forms handling time and expense inherent in producing a complete price list. As the result, most sales organizations reissue their price lists once a month or less frequently, relying upon individual price change bulletins to notify sales persons of any revisions that are to be hand-posted to the price lists and that will remain in effect until the next revision of the price lists. COM, as the result of its high throughput speed, and the speed and low-cost with which duplicate microform-based price lists can be produced, has found ready acceptance among sales organizations as the media in which all price lists will be issued. The COM process's speed and low costs also makes it possible to reissue price lists whenever they are revised, thus reducing the likelihood that individual salespersons will neglect to hand-post accurately and immediately upon being notified.

The foregoing are typical of the many ways organizations of every size and type are utilizing COM as an alternative to the computer output printer. Figure 9-1 is a listing of the additional applications in which this direct replacement is occurring.

COM as a Method for Preparing Business Graphics

The use of a COM device that has both an alpha-numeric and a graphic output affords organizations the opportunity to simplify and reduce the costs of preparing and distributing many of the periodic operational, performance, and status reports they now must generate either manually or in a combined manual and computerized application. The following applications will indicate just how effective this use of COM can be.

Preparation and Updating of PERT Schedules

A consulting architectural firm that must plan for and coordinate the services and performances of a variety of construction crafts, designers, and office furnishers

Accounting and Finance

o Accounts Receivable History File
o Accounts Receivable Invoices
o Monthly Aging of Accounts Receivable
o Accounts Payable Listings
o Check Registers
o Trial Balances
o Claims Registers
o Payroll Registers
o General Ledgers
o Invoices
o Fixed Asset Listings
o Accounting Code Master File
o Budgets

Personnel and Industrial Relations

o Personnel Rosters
o Telephone Directories
o Seniority Listings
o Pension Ledgers
o Management Development Files

Purchasing

o Vendor Master Listing
o Merchandise Control Reports
o Buyer Analyses Reports
o Purchase Order Status Reports

Sales and Marketing

o Price Lists
o Customer Name and Address Files
o Sales Reports
o Market Analyses Reports
o Order Status Reports

Production

o Inventory Management Reports
o Standard Cost Summaries
o Labor Control Reports
o Performance Reports/Production Statistics

Figure 9-1   Applications—COM as a direct replacement for output printers.

has found that the PERT chart provides an excellent means of both scheduling the overall project and updating the schedule to reflect completed tasks or the modification of completion dates. Many years ago, a software house offered a commercial software package for using PERT for sale, and this firm was one of the first to purchase and install this package. As a result, they were able to maintain their various tasks and their status and estimated completion dates on computer—a great help in updating these task lists. However, the depiction of these task lists and completion dates in a graphic form required the manual preparation of the PERT chart. The advent of COM devices with both alpha-numeric and graphic capabilities changed all that. Now this firm has developed a computer-based application in which the digitized data reflecting the task lists and completion dates are produced, via COM, in the form of an annotated PERT chart.

Preparation of Financial Reports

A manufacturer of communications equipment utilizes a COM device with both alpha-numeric and graphic capabilities to produce financial reports for its senior management. Their computer program has been developed so that the various sales, manufacturing, and other data appear as both statistical tables and graphs, which are then converted to a COM format, duplicated, and distributed in a COM microformat to the various recipients.

COM as a Source for Producing Quality Graphics

A number of farm and construction equipment manufacturers utilize COM devices with graphic capability for the production of new and revised engineering drawings. Programmatically, the computer produces the drawing outputting its image of a CRT-based visual display. Using a light-pen, the engineer or draftsman may make any necessary modifications to the drawing. He then reenters the revised drawing into the computer's memory, from which it is recalled, projected on a suitable COM device's display unit, and microfilmed. The microform containing the engineering drawings is then duplicated and used for active reference.

ESTIMATING THE COSTS AND SAVINGS OF A COM APPLICATION

Put into proper perspective, COM is simply another way of preparing and distributing computer-generated forms, reports, and other output. Therefore, to assure its economical and effective use, it is necessary that its costs be compared with those involved in the preparation and use of computer-generated hard copy output, in a given application, to estimate the costs of each and the cost savings that will accrue from either conversion to COM or the continuation of the hard copy-based application.

Estimating Hard Copy Costs

There are two alternatives to estimating the costs of a computer-generated hard copy output-based system:

• **Standard Performance Data.** A method that is well suited for estimating the costs of a *planned* but not yet implemented system. It is also usable for estimating the cost of any ongoing, hard copy-based system for which no historical cost and performance data is available.

• **Historical Data.** A method that uses the actual cost and performance records maintained by the user to derive estimates of future costs. Naturally, this method is applicable only for ongoing systems for which the user has accumulated cost and operational data.

Estimating Costs Using Standard Performance Data

With this method, we must estimate *material costs* and *computer printing and forms handling costs, distribution* and *filing costs.*

Material Costs

Included in material costs are:

• stock or custom designed computer forms,
• binders (if any), in which the forms will be filed.

These costs can best be estimated by requesting your organization's Purchasing Department to obtain indicative price quotations from one of their vendors for supplying the estimated annual quantities of forms and binders required. In determining the number of forms required, it is advisable to assume a total of 64 lines of data will be printed on each page or form.

Computer Printing and Forms Handling Costs

These costs include the operating costs (equipment, labor, and overhead) involved in the computer printing and forms handling (the bursting, decollating, binding) of the forms prior to distribution.

The author's experience indicates that these costs can be realistically estimated based upon a percentage of the computer's rated printing speed. As the result of time lost while setting up the computer for operation, and in the various forms handling operations, the computer's rated printing speed will be reduced.

The greater the number of copies of a given form that will be produced by the computer, the greater will be the loss in rated printing speed, as indicated by the following table.

Number of Copies in Form Set	Percentage that Rated Printing Speed will be Reduced
1	25%
2	30
3	35
4	40
5	45
6	50

The following procedure may be used to estimate the net printing speed and the costs of computer printing and forms handling:

- Estimate the number of forms that will be produced annually.
- Then determine the number of copies that will be contained in each of these forms.
- Referring to the above table for the number of copies in the form set, multiply the computer's rated printing speed by the percentage that the rated printing speed will be reduced. This will indicate the estimated lines per minute of reduction anticipated.
- Now calculate *net* printing speed by subtracting from the rated printing speed the number of lines per minute of reduction anticipated.
- Divide this answer by 64 to obtain the number of pages, or forms, that will be produced in a minute. Multiply this result by 60 to convert to hourly output.
- Divide estimated number of forms to be produced annually by hourly output to estimate the number of hours of computer and forms handling time required annually.
- Multiply this answer by $100 per hour to obtain the estimated annual cost of computer printing and forms handling.

Distribution Costs

Distribution costs may best be estimated by simulating the various tasks involved in wrapping, addressing, mailing, or delivering the computer-generated forms or reports to the several addresses. It is recommended that a prototype computer run be prepared, containing an average number of pages (or forms), and that this be subjected to each of the distribution processing steps that the actual pages would be. By this method, it will be possible to estimate the various times and material costs that would be involved in each of the distribution processing steps.

Storage Costs

Storage costs can be realistically estimated by assuming that:

- office storage costs (which include costs of filing equipment, floor space, and personnel) will average $10.00 per linear foot of records per year;
- inactive records storage costs (which include the costs of transportation to storage, filing equipment, floor space, personnel and destruction costs) will average $2.50 per linear foot of records per year;
- the typical computer-generated output will contain 1,800 pages per linear foot, and will be retained for an average of 18 months in office storage and an additional two years in inactive storage before being destroyed.

By applying the above standards, the costs involved in office and inactive storage will be estimated. Total costs of the proposed hard copy-based system can now be estimated by adding together the costs of materials, computer printing and forms handling, distribution, and storage.

Estimating Costs Using Historical Data

This method makes use of the cost and operational data already available within the organization; as such it is far more reflective of the true costs than is the previous method. Since the historical data costing method is concerned with an ongoing hard copy output computer application, the costs of materials, computer printing and forms handling, distribution, and filing will usually be readily available.

Material Costs

Material costs may be obtained by asking the Purchasing Department to research their records and determine:

- the volume of stock or custom designed forms and binders purchased for use in the application under study during the past year;
- the current costs of that volume of forms and binders.

Computer Printing and Forms Handling Costs

These costs can be ascertained by referencing the Data Processing Department's operating logs, determining the number of hours required to print the hard copy output as well as to decollate, burst, bind, and so on, the output prior to filing. After multiplying these time requirements by the current hourly computer and forms handling usage charges, one may accurately calculate these costs.

The above costs may be obtained or estimated in the following manner:

- *Preparation of a print tape* involves, depending upon the computer configuration involved, either the transfer (or unloading) of the data from disk to magnetic tape, or the reformatting of a magnetic tape from a standard computer acceptable format to one that is compatable with the COM device. This cost may be estimated by having the organization's Data Processing Department calculate the probable time that would be involved in either the disk to tape conversion or in the tape reformatting and then multiplying these totals by the hourly charge for this operation.
- *Service Bureau charges* are the actual charges imposed by the commercial Service Bureau for the COM conversion and preparation of distribution microforms. These costs must include, if applicable, such supplemental charges as that imposed for the forms overlay, and any other additional costs.
- *Distribution costs* are minimal with COM. A good rule of thumb is to figure these costs as averaging $1.00 per recipient for each cycle.
- *Filing costs* are also negligible. However, if readers or other micrographics hardware must be purchased as the result of the implementation of the COM-based application, the amortized value of such hardware must also be calculated.
- *Distribution and filing costs* can best be obtained by determining the costs involved in the distribution and filing of a single processing cycle's hard copy output, and then multiplying these costs by the number of cycles that the application will be run annually.

If one were to add all these costs together, a reasonably accurate figure will be available for the total costs involved in producing the output in a hard copy format.

Estimating Costs of Service Bureau-prepared COM Output

If the COM output will be prepared by a commercial Service Bureau, the determination of the estimated costs involve the following costs:

- Preparation of a print tape.
- The Service Bureau's charges for creating the COM master and making the required number of distribution copies from that master.
- Distribution costs.

- Filing costs, including the costs of such micrographics hardware as may be acquired for use with the application under study.
- If in addition to the COM output, hard copy will also be produced, the attendant costs involved in the production, distribution, and filing of that hard copy output must also be calculated.
- Hard copy costs, if any, should be determined by either the standard performance data or historical data method, as applicable.

The above cost factors will afford a reasonable estimate of the costs that would be incurred were the application under study to produce a COM-output using the capabilities and facilities of a COM commercial Service Bureau.

Estimating Costs of In-House Produced COM Output

The following costs will be involved in the preparation of the COM output in-house:

- Hardware amortization costs.
- Microfilm stock for COM master and distribution copies.
- Film processing and duplicating costs.
- Distribution costs.
- Filing costs.
- As in the case of commercial Service Bureau-produced COM, if hard copy output will also be prepared all attendant costs involved must also be determined.

Hardware amortization costs

These costs may be estimated by determining the monthly amortization of the COM device, film processor, duplicators, readers, and other hardware that will be used in the application, and then prorating this cost based upon the percentage of total usage that the application under study represents.

Microfilm Stock Costs

These may be determined by multiplying the quantity of master and duplicate microforms required annually by the cost of such microforms.

Film Processing Costs and Duplicating

These costs may be estimated by sampling the time required to process a COM master microform (if applicable) and to duplicate that master to produce the required number of distribution copies, then extending those times to reflect the total number of masters and duplicates involved. This latter figure may then be multiplied by the average wage paid the organization's micrographics technicians, plus fringes, to estimate film processing and duplicating costs.

Distribution, Filing, and so on

The costs of any additional hard copy output may be calculated as described above.

Again, by totalling these various cost factors, one will have a reasonably accurate estimate of the costs involved in generating an in-house produced COM output. It then becomes a straightforward task to array the costs of the hard copy, COM, and combination hard copy/COM applications to determine which will prove the most economical.

10

Micrographics and the "Office of the Future"

In the preceding chapters, we have examined, in considerable detail, how micrographics may be used to reduce operating costs and improve overall efficiency in *today's* offices. But what of the future? What role will micrographics play in *the office of the future*; that automated, largely paperless operating environment in which virtually every phase of information management, from creation to distribution, through processing, storage, and retrieval, will be electronically handled and controlled? Will micrographics revert to its former position as a means of economically maintaining large volumes of infrequently referenced information, or will it continue to be a viable alternative to other recordkeeping media and an important technique for use in active information storage and retrieval applications?

SCENARIO OF THE OFFICE OF THE FUTURE

Before exploring this question, however, let us first examine how information will likely be processed in the office of the future. It appears, based upon evolving technologies and the predictions of Office Futurists (persons whose primary function is to plan how their organizations will be operated in the future), that the following scenario will be involved.

Hard copy records will be virtually nonexistent. The greater majority of data will be created without hard copy output by such electronic means as voice input, optical character recognition, or word processors and will be stored, pending recall, in either a machine-processible, digitized format, or in a microfilm with some means of automated retrieval. This information would be retrieved upon demand and within seconds be displayed on a conveniently located CRT device. After the data had been reviewed and the reference had been completed, the information would then be revised or updated, as required, by the person who recalled the information from storage. This revision or updating would be performed by means of the CRT's keyboard, after which the information would be reentered into the computer or other applicable system, pending subsequent reference.

Information will be distributed by means of teleprocessing and/or telecommunications networks and will be delivered to addressees in such varied formats as facsimile-produced microforms or visual displays. Again, hard copy will be the exception.

MICROFILM'S ROLE

Most Office Futurists agree that due to its various characteristics and capabilities, such as its light weight, the ability to store many records in a small area, ease of reproducibility, and the like, microfilm will play a very significant role in the office of the future.

The general consensus, which is supported by a variety of product lines recently introduced by both micrographics manufacturers and nonmicrographics manufacturers alike, is that micrographics will be one of a group of interfacing technologies—along with data processing, word processing, and telecommunications—that will be integrated to form the basis for a system that combines the features of two or more of the interfacing technologies. We will therefore have a system that integrates computers, facsimile transmission, and micrographics, or possibly word processing, optical character recognition, and micrographics.

As the result, we will have a wide range of possible systems in which one may compensate for the limitations of one technology by the introduction of another of the interfacing technologies. The resultant system will, therefore, be more efficient and effective than would be a system that includes just one of the interfacing technologies.

There are a number of products both on the market and under development that are based upon this interfacing technologies concept. It will provide insight into the systems of the office of the future if we were to examine these.

COMPUTER-INPUT-MICROFILM (CIM)

Data processors have long sought cost-effective ways of coping with what they commonly refer to as "the input bottleneck"—their inability to feed data into computers at speeds that even begin to approximate the computer's ability to process and output that data. Today, as in the past, data entry is a slow, labor-intensive, error-prone manual operation. Clerks still spend their working days encoding forms for data entry, and keypunch and other data entry operators still manually convert that encoded data to a computer-processible format and language. Consequently, the costs, time, and error factors associated with the inputting of data are generally, according to a U.S. government survey, greater than the combined total of those same factors in the computer processing and output phases.

Optical character recognition (which is more commonly refered to as *optical scanning*), a data entry technique that utilizes an electronic speed-reading machine called a scanner to read and identify ordinary alpha-numeric characters, marks, and symbols and almost instantaneously converts them into a computer-processible code, offered a potential solution to the input bottleneck. Such equipment could read, identify, and translate as many as 3,700 full lines of information into computer-processible code in a single hour—the output of 150 *skilled* keypunch operators. However, optical scanning has achieved only limited acceptance among data processing professionals who found that its operating requirements, in terms of the source document's format and quality, was just too restrictive to permit its use in any application in which the preparation of the source document could not be rigidly controlled. There is little dispute among both users and nonusers, however, that in the correct application, optical scanning will provide an economical, efficient alternative to keyboard-based data entry.

The mating of the micrographics and optical scanning technologies to form an entirely new data entry technique known as computer-input-microfilm (CIM) provides a means of overcoming the optical scanner's aforementioned source document format and quality limitations and of better coping with the input bottleneck.

Although several manufacturers have been working on the development of CIM systems, and several have been brought on the market, their cost has made them prohibitive for all but the largest organizations. FOSDIC, one of the first of the CIM systems, which was developed by the U.S. Weather Bureau, is one such system.

The Weather Bureau had a library of more than 300,000 tabulating cards, each of which contained meteorological data used on a continuing basis. The volume of such cards was increasing at a rate of over 10% per year. Additionally, after several years of continuous use, the tab cards would wear out, and new ones would have to be key punched as their replacements. As described before, the ever-present input bottleneck was creating operational and scheduling problems for the Bureau's Data Processors.

FOSDIC (mnomic for *f*ilm *o*ptical *s*ensing *d*evice for *i*nput to *c*omputers) was developed to correct these problems and to bring the burgeoning files resulting from the creation of these tabulating cards under control. With the FOSDIC CIM System, each tabulating card was photographed, in a cine mode, by a high-speed microfilm camera equipped with an anomorphic lens—one whose use results in a slightly distorted image. The resultant microimage of the tabulating cards is "stretched" so that it appears longer than normal and the punched holes appear to be almost square. This permits more microimages to be fitted on the micro-form and facilitates the optical scanner's electronic beam's sensing of the holes.

Computer-input-microfilm holds great promise for the future, not only as a means of reducing the input bottleneck, but also as an alternative to magnetic tape as a data storage and input media. Despite advances made in magnetic tape technology of late, the density of microforms still compare favorably or exceed that of even the most densely packed magnetic tapes. In those applications, therefore, in which the reentry of computer-generated data occurs with infrequent regularity, there are significant cost reductions possible if the data were main-tained in a CIM format rather than on magnetic tape.

MICRO-FACSIMILE

Micro-facsimile is a developing technology that merges micrographics with fac-simile transmission to reduce the time and costs involved in transferring informa-tion from one location to another. Micro-facsimile systems provide for the digital scanning and subsequent transfer of microimages, in either aperture card, con-tinuous roll, or microfiche format, over telephone lines or via satellite or micro-wave to a facsimile receiver. Upon receipt, the digitized information is converted to ordinary alpha-numerics and is either printed out in a hard copy format or is displayed on a CRT device for the recipient's reference and use. Although it is not yet commercially available, a third alternative form of output—a microform—is being actively worked on by several facsimile manufacturers.

The impact of micro-facsimile on manual information storage, retrieval, and distribution is significant. Such a concept will make feasible the establishment of large data bases, stored in both computer-processible and microfilm formats, that would serve the needs of all functions of the organization, regardless of their location, via telecommunication links rather than ordinary mail or messenger services. With micro-facsimile, active data could be stored on magnetic tape or disk, be updated as required, printed out in a microformat and be transferred, via facsimile transmission techniques, to users located anywhere in the world, within a matter of minutes. The time, cost, and possible loss or damage of the records that is inherent in a standard hard copy or microfilm-based information transfer, would no longer exist.

MICROPUBLISHING

As the costs involved in the printing, distributing, and maintaining of new, reformatted and previously published books, periodicals, catalogs, manuals, reports, and other similar multipaged records continue to rise, it may be anticipated that micropublishing—the production of such printed materials in micrographic form—will also increase.

There are three distinct steps involved in the micropublishing process:

- Filming of the printed materials using either a rotary, planetary, or step-and-repeat microfilm camera.
- Processing of the exposed microfilm and the creation of a microform master in either a continuous roll or a microfiche format.
- Duplication of the microform master using either a roll-to-roll or fiche-to-fiche duplicator to produce the additional copies required for distribution purposes.

Micropublications show every indication of becoming the mainstay of the business and governmental libraries of the future, affording an economical method of:

- acquiring rare and out-of-print books,
- maintaining collections of bulky periodicals,
- reducing damage to the printed materials, and thus eliminating the need for their repair, rebinding, and refurbishing.

Reliable industry sources project a five-year growth rate of over 20% annually for micropublishing, due, in a large part to such factors as:

- the progression of microfilm technology to the point where the quality of the microimage approximates that of the printed original,
- microfilm readers and reader-printers have reached a uniformly high level of performance and operational simplicity, resulting in the elimination of what formerly was a major user objection to micropublishing applications,
- organizations of every type, faced with spiraling operating costs must seek new methods of operation in order to maintain a competitive position in their markets (this has resulted in many organizations taking a second look at micropublishing and, in view of the steadily-increasing costs involved in the printing, distribution, and maintenance of hard copy, deciding to convert to a micropublishing format),

- possibly most important, a whole new generation of office personnel is functioning today, a generation that has grown up with the computer and microfilm and is less emotionally and psychologically tied to paper than preceding generations were.

Micropublishing will play a major role in the office of the future, as organizations seek new, more efficient, space and labor saving methods of maintaining large volumes of printed information in their immediate working areas.

WORD PROCESSING

Word processing—the automated production of typewritten materials—is generally considered to be the cornerstone of the office of the future, providing a rapid, cost-effective alternative to manually prepared letters, reports, forms, and other typewritten documents. Reduced to basics, word processing utilizes a programmable typewriter that is equipped with a memory, and contains a variety of options such as a file maintenance or a communications capability. With such word processing equipment, materials are typed in draft form, and as the hard copy is prepared, an exact duplicate of that document is stored and indexed in the word processor's memory. After review and correction or modification by the originator, the draft is "played-out" from memory and the necessary revisions made in their exact positions by the word processing operator. When the draft has been updated, the operator may then produce the final copy, automatically and error-free, at speeds of upward of 140 words per minute.

The interfacing of word processing and micrographics technologies in the office of the future will likely be in two areas:

- *Electronic mail,* in which correspondence and other typewritten documents will be produced by word processor, and after conversion to a suitable microform, be transmitted via micro-facsimile to the addressee's location. There it will be printed out in either hard copy or in a microform, for distribution, or else be input into a CRT-accessed data base.
- *Data base intermediary,* in which the microform serves as an intermediary between the word processor and the computer. Documents produced by a word processor would be reformatted into a computer-processable format, and then be input into a COM device. The latter will produce as its output, a continuous roll microform or a microfiche, in which all data is printed in an optical scanner-processable type font. When it is necessary to update the word processing output, the microform would be inserted into an optical scanner and be reentered into the word processor's memory in a reversal of

the output procedure described above. After the updating has been completed, the reformatting and COM input will again occur, and an updated microform will be produced. Should retrieval and reference be required, the COM-generated microform would be removed and be:

- read and (if applicable) copied by a suitable micrographics reader or reader-printer,
- transmitted via micro-facsimile to the requestor's location where it will be formatted for use and reference, or
- input into a computer to produce hard copy output.

COMPUTERIZED MICROGRAPHICS

As mentioned in Chapter 5, Automated Document Storage And Retrieval (ADSTAR) Systems link the computer and micrographic technologies to effect systems that enable large volumes of documents to be indexed, stored, and retrieved quickly and accurately by any number of different parameters. In ADSTAR Systems, the computer (or in many cases, the minicomputer) provides for rapid, in-depth indexing while the micrographics technology affords high storage density at low cost and assures a means of economical and ready duplication and reproduction.

ADSTAR Systems typically provide for the random microfilming of source document and the subsequent random storage of the resultant microimages and their associated indexing parameters. The microimages are stored within the ADSTAR System in a variety of microforms; the indexing parameters and the storage address of the microimages are stored within the ADSTAR System's memory (often on a floppy disk). When retrieval is required, the applicable search parameters are input to the system by means of a CRT console. The system searches its memory comparing the search parameters to those of each microimage. When a match is found, the storage address is noted and the system is instructed to retrieve and display the applicable microimage on the CRT. Virtually all ADSTAR Systems provide a hard copy print-out option, should the user require a reproduction of a given microimage.

ADSTAR Systems are finding increasing acceptance among governmental, industrial, financial, and nonprofit organizations of every type, as a means of maintaining large, subject-oriented records accumulations. Most industry observors feel confident, according to a recent survey by the author, that ADSTAR Systems will, along with word processing, provide the cornerstone for the office of the future, providing a cost-effective means of storing and retrieving both active and inactive documents.

11

The Micrographics
Management Program

It is the writer's observation, gained from assisting hundreds of organizations located in the United States, Canada, and Europe establish cost-effective micrographic systems, that the average user is not in a position to exploit his organization's micrographics resources to their fullest potential. To understand the basis for this statement, let us consider the evolution of micrographic systems in the typical organization. One will find that with few exceptions their growth was not planned; like Topsy, micrographic systems just grew. Shortly after World War II, the organization probably began filming cancelled checks and accounts payable files in order to recover space and filing equipment. Just as probably, this microfilming was done by the Central Files personnel utilizing a rotary camera, to produce a flash indexed negative roll film.

When aperture card-based engineering drawing systems came into being, it is probable that such systems were developed by the Engineering or Drafting Departments and that the microfilm cameras, readers, duplicators, and other hardware supporting this system were bought without consideration being given to any existing or planned micrographic hardware.

The same situation is generally true of microfilm jackets, microfiche, and COM-based applications. In the typical organization, each was generally installed to meet a specific, immediate need, and each is probably administered by a different organizational unit.

Indexing is probably nonstandard, based upon the requirements of each individual application. The hardware involved was likely procured specifically for the application at hand.

Consequently, the average organization's micrographics program is not a coordinated program at all. Rather, it is a fragmented, nonstandardized, often inefficient, typically poorly directed and managed approach to solving individual problems. Often, there will be unjustifiable duplication of hardware, such as two rotary cameras being used in adjacent departments when the combined workload indicates that only one is actually required. Various special-purpose readers and reader-printers will be in use, each procedured to satisfy a unique application. The lack of standardization regarding indexing, microform selection, and reduction ratios will generally severely impede the interchangeability of microfilmed records and well may preclude the use of more efficient, less costly hardware. Also, and perhaps most important, where no central control has been established to appraise new and revised micrographics applications, there is a decided probability that systems that cannot be cost and operationally justified will be implemented and that users will purchase more elaborate, more expensive hardware than their retrieval requirements justify.

To avoid these pitfalls and to place an organization in a position where it can readily exploit micrographics' operating and cost reduction potentials, a Micrographics Management Program should be developed and administered. Such a program will provide for the application of systematic controls over (a) the development, implementation, and maintenance of micrographic systems, and (b) the standardization, procurement, and utilization of microfilm cameras, readers, and other hardware.

Such a program will centralize the various segments of micrographics. Responsibility for micrographics management, rather than being decentralized among its primary users (many of whom may lack the technical knowledge to properly plan for its cost-effective use), will be centralized in a staff department whose sole function is to achieve maximum cost and operational benefits from the use of micrographic systems.

Organizational Placement

To be effective, the Micrographics Management Program must be autonomous. It should not be made a part of an existing Records Management Program, nor of a Reprographics Management Program, for example, although it must certainly interface with those programs. Ideally, a new department should be created to develop and administer the Micrographics Management Program—the Micrographics Management Department. This new department should be headed by a Micrographics Manager, who would be on an organizational level comparable to that of the organization's Records Manager, Forms Manager, or its Reprographics Manager.

Organizational placement, while important, is secondary to the amount of support the program receives. Micrographics Management is, after all, a control program, and it is human to resent and resist being controlled. Therefore, strong management is definitely required to assure that people are not able to ignore or circumvent the program's procedures. For this reason, the Micrographics Management Program should be assigned to a staff level executive with a proven interest in cost consciousness and a record of aggressiveness where cost controls are concerned.

Responsibility for Program Installation

Unless management's decision is to implement the new Micrographics Management Program in as short a time as is possible, there is no reason why it cannot be installed by the organization's own personnel. Micrographics management concepts and techniques are not complex, and may be mastered by one who makes a concerted effort to do so. This book provides the information one requires to establish and administer a Micrographics Management Program. By constantly referring to this book as problems arise, and by adding to and upgrading knowledge through the sources to be described shortly, nearly anyone can obtain the knowledge required to administer an efficient, cost-effective Micrographics Management Program.

Definite benefits, however, will accrue to an organization that decides to use the services of a knowledgeable consultant, provided that the consultant's services are properly utilized. E. R. Squibb & Sons, Inc., is an organization that decided to utilize a consultant as a means of quickly installing their Micrographics Management Program and achieving the operational and cost benefits inherent in such a program far sooner than would be possible were the installation to be accomplished by internal personnel who had no prior experience in the establishment and installation of such a program or in the planning for its continued administration. Squibb's management was very careful to assure certain conditions were met before contracting for the consultant's services:

- First, they were careful to select a consultant who included among his staff individuals who had extensive "hands-on" experience in micrographics.
- Second, they were careful to specify that these individuals were to be assigned to their project, and that any other of the consultant's personnel could be added to the Project Staff only with the explicit approval of E.R. Squibb & Sons, Inc.
- Third, E. R. Squibb & Sons, Inc., included in the consultant's contract the stipulation that at least two Squibb employees will work along with the consultant in developing and installing the Micrographics Management Program. In this way, they will obtain valuable on-the-job training and be in a position to assume the program's management *after the consultants leave.*

- Fourth, the consultant was required as a deliverable item under his contract, to prepare an operating manual for Squibb's continuing use. This manual establishes the Micrographics Management Program's operating procedures, specifies indexing standards and guildlines, details the criteria for microform selection, and so forth. Such a manual, Squibb found, added to the contract's cost, but the resultant benefits more than offset the additional expense.

Staffing the Micrographics Department

The staffing of the new Micrographics Management Department is complicated by two major situations:

- With the exception of a limited number of univcisity extension courses, no formal education training is offered at either the college or high school level to help persons prepare for careers as Micrographic Technicians, Micrographics Systems Analysts, or Micrographics Managers. Most micrographics professionals became involved in this field by necessity, rather than by plan. As they saw an application for cost saving through the application of micrographics in their primary occupations, such persons were forced to acquire the technical knowledge to be able to design, implement, and administer the micrographics application. Thus, we find many computer professionals are also knowledgeable about COM; many draftsmen and engineers are highly skilled in engineering micrographics, and such varied occupations as accountants, records managers, methods analysts, librarians, systems analysts, and information specialists will possess a formidable knowledge with respect to the business applications of source document micrographics.
- As the result, it is unlikely that any one individual in the organization will possess the skills necessary to implement, in the near term, a broad-based micrographics program that incorporates the full range of source document and COM applications.

Recognizing these factors, the author suggests that the selection of the Micrographics Manager be based upon *potential* rather than *current* ability. By selecting one who possesses proven analytical ability, is a self-starter capable of functioning independently, is a good communicator both orally and in writing, and has the interest to upgrade knowledge through what will primarily be after-hours efforts, the organization will solve the dilemma posed by the lack of personnel trained and experienced in all functions of micrographics.

Individuals chosen for positions below the Micrographics Manager's level should be on the following bases:

- **Micrographics Systems Analyst.** There are five basic attributes that the prospective Micrographics Systems Analyst must possess, if he is to develop into an effective operative:

  - **An Inquisitive Nature.** This is an essential quality for any type of analytical work. The analyst must constantly question the need for each procedural step, each indexing method, each item of hardware, in an effort to uncover a more economical, more efficient, more effective alternative. Unless the analyst possesses a basically inquisitive nature, countless opportunities for systems improvements and cost reductions are likely to slip by.

  - **A liking for detail work.** The design of a micrographics-based system is, pure and simple, detail work. The better the analyst is able to recognize and plan for all the possible contingencies and variations that could occur the less the likelihood that the system will encounter stoppages as the result of (for example) inadequate depth of coding, improper hardware, inability to interchange microforms between two or more locations, and the like.

  - **Tact.** The primary purpose of Micrographics Management Programs is assuring the cost-effective use of micrographics in the organization's operations. This requires the Micrographics Systems Analyst to turn down certain user requests and recommendations when they cannot be cost or operationally justified and to suggest alternatives for other user requests and recommendations when cost and operational benefits will result. The analyst must decline and modify *tactfully,* if the user is not to become antagonized and possibly conclude that the program is hindering rather than assisting him perform his responsibilities.

  - **Ability to reach decisions without complete facts.** The effective analyst will often be required to estimate what might happen in a given situation were one or a number of possibilities to occur. Therefore, the potential Micrographics Systems Analyst must be able to make decisions when only partial facts, or an estimate of the facts, are available.

  - **Good oral and written expression.** Since most micrographics systems analyses will require both a written report of findings and recommendations and a verbal presentation to the using departments the analyst must be able to clearly and accurately express himself both orally and in writing.

  - **Micrographics Technician.** The Micrographic Technician's primary function will be to operate the rotary and planetary camera, COM recorders, film processors, inspection equipment, duplicators, and the like that are involved in the preparation of original and duplicate microforms. The skills required for the operation of such hardware may be acquired by virtually any individual who makes a conscious effort to apply himself. To reduce the likelihood that the Micrographics Technician will not operate such hardware in an efficient, economical, correct manner, it is suggested that potential technicians be limited to those having the following three characteristics:

- **Ability to comprehend what he reads.** The Micrographics Technician will constantly reference the various equipment operating manuals and internal procedures developed by the Micrographics Manager and the Micrographics Systems Analysts. It is essential, therefore, that the technician have the ability to read, comprehend, and follow the instructions contained in such manuals and procedures.

- **Alertness.** Since he will be working with mechanical equipment that may malfunction at any time, the Micrographics Technician must be alert to the various warnings and symptoms of such malfunctions and take corrective action.

- **Attentiveness to detail.** The production of a usable microform is largely determined by how well the technician follows the equipment manufacturers' operating procedures for the cameras, duplicators, and other hardware involved, as well as the Micrographics Systems Analyst's instructions for indexing, formatting, and duplicating the microimages. Carelessness or inattentiveness on the Micrographic Technician's part can well lead to an omission or error that will severely effect the quality or usability of the microfilm-based records.

Recruit from Within or Hire from Outside?

The decision to recruit micrographics management personnel from within the organization or to hire experienced persons from the outside must be made on the basis of how much time the organization's management is willing to devote to getting the program fully operative.

Typically, from nine months to one year will be required for a neophyte to gain the technical skills and operating experience necessary to become proficient in his duties. At the end of that time, such individuals will probably have acquired the skills necessary to administer and operate an ongoing Micrographics Management Program. If however, the organization's management is unable or unwilling to fund the Micrographics Management Program for such a nine to twelve month period, there is no alternative but to hire experienced persons from outside the organization.

Training and Upgrading of Micrographics Personnel

There are, basically, four sources for obtaining and upgrading micrographics skills:

- Training courses offered by various associations, educational institutions, and professional trainers.
- On-the-job training under the tutelage of a micrographics consultant, a more experienced micrographics employee, or a manufacturer's representative.
- Selected readings.
- Attendance at trade shows and exhibits.

The training courses that are offered by a relatively small number of universities and colleges are generally broad in scope, and provide an overview of micrographics, its applications, limitations, and advantages, rather than becoming deeply engrossed in the technology itself.

The same is true of the course offerings of such professional associations and trainers as the American Management Association, Executive Enterprises, CEIR, and AMR, Inc.

The main advantages that such formal training affords are a structured approach that insures that the attendee will obtain an understanding and appreciation of microfilm-based systems and suggestions as to how such systems could be utilized by his own organization. Additionally, the contacts made at such courses with others is extremely beneficial over the long term.

On-the-job training involves actual hands-on instruction in microfilm technology and systems design supervised by a more knowledgeable individual, such as a consultant, a manufacturer's representative or an experienced member of the Micrographics Management Department, if one is available. Such training is tutorial in approach. The "trainee" observes the "trainer" as he (for example) operates a microfilm processor. Then he operates the equipment himself, while the "trainer" critiques his performance. This process of observation, imitation, and critique may be utilized with any facet of micrographics—from the development of a COM application to the operation of an item of hardware, to the modification of ongoing micrographics systems. It presupposes that a qualified "trainer" is available either within the organization or that the services of an outside "trainer" can be procured. If such is not the case, the on-the-job training method will not prove feasible.

There are a number of business magazines that contain one or more articles, columns, or features each month (on the average) on subjects of interest to the micrographics professional. Such articles will prove extremely valuable to both the seasoned micrographics person and the neophyte. For the experienced person, they provide a convenient method for staying abreast of new developments, equipment, and applications. For the neophyte, they provide a means for broadening knowledge of contemporary micrographics. Among the publications that every micrographics personnel should subscribe to are the following:

- *Graphics Arts Monthly*
  Dun Donnelly Publishing Company, publisher
  666 Fifth Avenue
  New York, N.Y.
- *Information and Records Management Magazine*
  and *Microfilm Techniques Magazine*
  PTN Publishing Corp., publisher
  250 Fulton Street
  Hempstead, N.Y.
- *Administrative Management Magazine*
  Geyer-McAllister Co., publisher
  51 Madison Avenue
  New York, N.Y.
- *THE OFFICE Magazine*
  Office Publications, Inc., publisher
  200 Summer Street
  Stamford, Conn.
- *Modern Office Procedures*
  Penton/ IPC, publisher
  1111 Chester Avenue
  Cleveland, Ohio, 44114

Graphic Arts Monthly, Information and Records Management, Microfilm Techniques, and *Modern Office Procedures* are "controlled circulation" magazines, available at no charge to persons engaged in such office administration activities as micrographics. The other publications, *Administrative Management* and *THE OFFICE* are available on an annual paid subscription basis.

Attendance at trade shows and exhibits, sponsored by such organizations as the National Micrographics Association and the Business Equipment Manufacturers Association, affords an excellent opportunity to examine the various micrographics hardware and to learn of new technological developments in the micrographics field.

Naturally, no single one of the above sources would prove an adequate means of training and upgrading the skills of micrographics personnel. A combination of each is essential to that end.

ANNOUNCING THE MICROGRAPHICS MANAGEMENT PROGRAM

After the Micrographics Management Program has been organized and staffed, it can be announced to the organization at large, and the initial phase of its implementation—the inventory of current micrographics applications and hardware—may begin.

It is suggested that this announcement take the form of a memorandum from the organization's Chief Operating Officer to all Department Heads. This announcement letter should accomplish the following:

- Announce the establishment of the Micrographics Management Program.
- Specify its program concepts.
- Outline the goals and objectives of the program.
- Assign responsibility for its development and administration to a designated Micrographics Manager.
- Explain that in the near future, this individual will contact each department to obtain further information regarding their current and projected micrographics applications and hardware.
- Indicate that the program has management's support and that its progress will personally be monitored by the Chief Operating Officer.

Shortly after the announcement has been distributed, the Micrographics Manager should issue a second memorandum. After referencing the Chief Operating Officer's announcement, this memorandum should request that each Department Head provide answers to the following questions within 10 working days:

- What micrographics applications are in use at the present time?
- What makes and models of micrographics hardware are used within the Department at the current time?
- Are any additional or new micrographics applications or hardware now under evaluation or scheduled for installation in the future. If so, which?
- Which formats are your department's microfilmed records maintained in? How are they indexed?

The Micrographics Manager should maintain a follow-up to assure all responses are received within the 10 day response period.

Analyzing and Summarizing Responses

By analyzing and summarizing the responses received from the department heads, the Micrographics Manager will be able to develop a concise profile of who utilizes micrographics within his organization, which records are maintained in a microfilm format, how they are indexed, and exactly what micrographics hardware is available within the organization.

Certain inefficiencies will readily be noticed as this profile is studied. One may find, for example, that inappropriate indexing is being used or that the microforms do not suit the application in which they are used. It may be observed that a department may have excessive micrographics hardware, or too little, or the wrong type. In all probability, numerous questions will be raised in Micrographics Manager's mind as to whether microfilm is the most efficient, most economical alternative.

Implementation of Improvements

The Micrographics Manager is now in a position to implement the services, guidelines, and standards that are required to assure efficient, economical, effective micrographic systems. Starting with the departments that make the greatest use of micrographics, he must examine each application and each item of micrographics hardware to determine:

- If the application can be cost and operationally justified.
- If the microform utilized is appropriate considering such factors as the type and size of the file, the frequency of reference, the need to create distribution and reference copies of both the entire microform and individual microimages.
- If the indexing employed is consistent with the number of microimages involved, the number of references and the number of parameters by which a given microimage may be referenced.
- If the readers, reader-printers, cameras, and other micrographics hardware employed by the department in the creation and use of the microforms are the most cost-effective.

Based upon the above, the Micrographics Manager may decide to recommend the elimination of certain micrographics applications, or the substitution of one microform for another (e.g., microfilm jackets for continuous roll microfilm, etc.), or to modify the methods of indexing, or to revise the hardware employed.

Such recommendations, of course, are subject to the Department Head's review and approval. However, if the Department Head disagrees with any recommendation, he should explain why the original method is preferable, and justify his preference in terms of operational or economic advantage. This will preclude the possibility that inefficient, uneconomical micrographics systems will continue unchallenged.

Once the individual micrographics applications are examined, the Micrographics Manager should turn to the broader questions of program continuity, namely:

- Which, if any, departments will continue to do their own microfilming, rather than having this service performed by the Micrographics Management Departments. In many organizations, the general practice is to allow Payroll and Personnel to utilize their own personnel for the filming of such confidential information as payroll checks, personnel jackets, and management development files, but to have all other records microfilmed by the Micrographics Management Department.
- What standards and guidelines should be instituted with respect to the selection of microforms and the utilization of indexing techniques to assure the interchangeability of such microforms, as required, between using departments;
- Will it be beneficial to standardize on specific readers, reader-printers, duplicators, and other micrographics hardware? If so, which should be selected? What disposition should be made of hardware now in the organization that is not the "standard"?

In conducting the analysis preparatory to the implementation of improvements, the micrographics professional will utilize the techniques recommended in Chapter 5. By doing so, the professional will assure that all salient points have been taken into consideration.

Reporting Program Accomplishments

The Micrographics Manager should tabulate any improvements that result from his efforts, assign a dollar value to the cost reduction or cost avoidance, and report such accomplishments to the Chief Operating Officer (or whoever sent out the program announcement) on a monthly basis. This will go a long way to establishing the value of the Micrographics Management Program to the organization, as well as increase the status of the Micrographics Manager and his staff.

Program Continuity

In addition to the study and improvement of existing micrographics applications, the Micrographics Manager must have the authority to review and approve or disapprove all proposals for the conversion of any hard copy or computer processible records to a microfilm format.

Such review and approval must extend to all hardware to be employed, all indexing systems to be used, all microforms to be created, and all quality control procedures to be followed.

It is essential, as a means of assuring such review, that the Micrographics Manager be notified early on, while it is still in the development stages, of any new or revised micrographics applications.

Glossary of Micrographics Terminology

Micrographics has a terminology of its own, which has largely evolved during the past three decades. Since an understanding of this terminology is essential to understanding the various literature dealing with micrographics, the following glossary of micrographics terminology, containing the most commonly used acronyms, technical jargon, abbreviations and terminology is presented.

Acetate Film Jackets. See Microfilm Jackets.

Alignment Targets. A guideline, generally a piece of white tape, that is placed upon a glass mask to permit rapid and accurate registration of copy on a camera bed. *Also*: a strip of white tape that is placed horizontally across the width of a reader or reader-printer's viewing screen to facilitate the reading of a line on a given microimage.

Alpha-Numerics. Letters of the alphabet, numerals, and special symbols (punctuation, ampersands, dollar signs, etc.).

Aperture Card. A standard 7⅜ × 3¼ inch tabulating card into which a rectangular hole or holes (or *aperture*) has been cut. The aperture(s) are specifically designed to hold one or more microimages.

Archival Quality. The ability of a microform to retain its image quality and to resist deterioration during a specified period of use and storage.

Aspect Ratio. The ratio of a given microimages' width to its height. This ratio may also be used to describe the ratio of width to height of an individual alpha-numeric character.

Automatic Aperture Card Mounter. Equipment that automatically cuts and counts microfilm in aperture cards.

Background. That portion of a microimage or of a paper record that contains no preprinted (fixed) or entered (variable) information.

Bit. A binary digit that is part of a two-digit numbering system that is standard in electronic data processing. Many of the automated retrieval systems make use of binary code in the indexing and retrieval of individual microimages.

Blow-back. See Magnification Ratio.

Blueprint. A full-size paper reproduction of a drawing or other document that has a white image area on a blue background.

Camera. See Rotary Camera; Planetary Camera; Step-and-Repeat Camera.

Card-to-Card Duplication. A contact printing process that will produce the microimage(s) contained in a unitized or a nonunitized microform.

Cartridge. A single core container enclosing roll microfilm designed for use with readers, reader-printers, or automated micrographics retrieval equipment.

Cassette. A double core container enclosing roll microfilm for use with readers, reader-printers, and retrieval devices.

Certificate of Authenticity. A legal form that is filmed as the last document of a records series and authenticates all documents preceding it as true copies of source documents for legal and/or regulatory purposes.

Cine Mode. A method of arranging microimages on a microform. With *cine mode,* individual microimages are arranged sequentially so that the top of one document is adjacent to the bottom of the preceding document.

Coding. A predetermined system for facilitating the location of desired microimages. Coding may take a variety of forms ranging from flash areas on roll microforms to binary coding in automated retrieval systems.

COM (acronym). Computer-Output-Microfilm.

Computer-Output-Microfilm (COM). A process in which computer-processible data, contained in magnetic tapes, disks, and so forth, are converted to ordinary alpha-numerics and produced on a microform without first being printed out as hard copy.

Comic Mode. A method of arranging microimages on a microform. With *comic mode,* individual microimages are arranged sequentially so that the right-hand side of the first microimage is adjacent to the left-hand side of the second.

CRT (acronym). Cathode ray tube.

Density. The degree of contrast between the image and the nonimage (background) areas of a microimage. Readily measured with a densitometer, density is generally expressed as a numerical equivalent.

Department of Defense Specifications (DoD Specs). Standards developed by the Department of Defense concerning microfilm camera work as well as microfilm processing, coding, and reproduction. All contractors and subcontractors submitting microfilmed records to the federal government under the terms of their contracts are required to conform to these specifications.

Descriptor. An indexing term that indicates something about a particular document, such as its subject, originator, and the like.

Developing. See Processing.

Diameter. The term that describes the number of times a document has been reduced or a microimage has been enlarged.

Diazo Film. A relatively slow print film composed of azo dyes that, in the presence of strong light and ammonia vapors, is capable of creating an image. Diazo provides a positive nonreversing image, with blacks being reproduced as blacks and whites as whites.

Document. Any form, report, photograph, drawing, correspondence, or other materials generated or received by an organization in the course of its business that records or transmits information.

Dry Process Film. Any film that is not developed by chemicals in solutions.

Dry Silver Film. A nongelatin silver film that is processed (developed) by the application of heat.

Duplicator. A device that makes single or multiple copies of a master silver halide microform.

Duo. A filming format that is created by passing the film through the microfilm camera twice and photographing across one-half of the film's width during each passage.

Duplex. A filming format that is created by the simultaneous photographing of both sides of a document and the positioning of these front and back images side-by-side, across the width of the microfilm.

EBR (acronym). Electron beam recorder.

Electron Beam Recorder. A COM device.

Electrostatic Printer. A reproduction process that utilizes static electricity for the formation of images.

Emulsion. A coating on a transparent film stock in which light sensitive materials have been suspended.

Enlarger. See Reader.

Enlarger-Printer. See Reader-Printer.

Fiche. See Microfiche.

File. Any collection of documents arranged in a predetermined manner, for example, alphabetically, numerically, or chronologically.

Film Chips. Roll microfilm that has been cut into small sections (or chips).

Film Jackets. See Microfilm Jackets.

Fixer. See Hypo.

Flash Card Indexing. A simple micrographics indexing system that subdivides a microfilm roll into smaller, more easily searched segments by inserting blank spaces at predetermined intervals. These blank spaces, when viewed on a reader or reader-printer, appear as white flashes.

Forms Overlay. A glass slide that is etched with the lines, captions and other features that make up the preprinted (fixed) portion of a form or report-form. This slide is placed onto the COM device's CRT, and the entered (variable) data appears on the face of the CRT. When the microimage is created both the forms overlay and the data appearing on the CRT are photographed, creating a form.

Frame. A single microfilm exposure.

Generation. A copy created from an original microimage or a copy that has been created from that microimage. Thus, the original microform would be the first generation film, while a copy of that microform would be a second generation film, and so forth.

Generation Test. A means of determining the number of times usable copies may be reproduced from succeeding generations of microfilm. In this test, copies are successively reproduced until an unreadable or otherwise unusable print has been generated. This indicates the anticipated range of copies that may reasonably be expected from that microfilm.

Grid. A pattern formed by the intersection of horizontal and vertical lines that are used to assign location designations to individual microimages contained in a microfiche or a microfilm jacket.

Hand Aperture Card Mounter. Hand operated equipment for cutting and mounting microfilm in aperture cards.

Hard Copy. The original document or reproduced paper copy made from a microform. *Also:* a term frequently used to describe the output of electronic data processing equipment.

Header Information. Indexing information, such as a description of the microimages' contents, and the like, that appears across the top width of a microfiche, ultrafiche or microfilm jacket.

Hit. A term used to describe the matching of items in a file that is being mechanically searched.

Hypo. Sodium thiosulphate, a chemical solution used in the processing of silver halide negative microforms to "fix" the images and render them impervious to future exposure to light.

Impact Printer. A mechanical printer in which the image is created by the striking of the printer's type against a form or other paper record.

Indexing. A description of the contents of a given microform or other record based upon such characteristics as subject matter, date, addressee or originator, and so on.

Intermediate. A microform, or a reproduction made from a microform, that is intended to serve as the master for making successive prints or reproductions.

Jacketed Microfilm. See Microfilm Jackets.

Latent Image. The invisible, undeveloped image that is created on a microform following exposure.

LBR. Acronym for laser beam recorder, a COM device.

LED. Acronyn for light emitting diode.

Magnification. The number of times a given microimage is enlarged for projection upon a viewing surface or when reproduced in a hard copy format.

Master Microform. The original, exposed microform from which subsequent reproductions will be prepared.

Microcard. See Micro-Opaque Card.

Microfiche. A unitized microform in which a group of related images is arranged on a transparent sheet of film in rows and columns, in the same manner as dates appear on a calendar.

Microform. A generic term for any form, usually film but just as well paper, that contains microimages.

Micrographics. The technology involved in the creation and use of microimages.

Microimage. A record that is too small to be read by the naked eye.

Micromation. See COM.

Micro-Opaque Card. A unitized microform in which positive microimages are arranged on one or both side of a white, opaque card, in the same manner as dates appear on a calendar.

Microphotography. The technology involved in the production of miniaturized film copies of documents.

Micropublishing. The printing of textual matter in a microfilm format, in lieu of or in addition to a paper format.

Microrepublishing. The printing in a microfilm format of textual materials that have previously been published in a paper format.

Negative Image. A photographic image in which the tonal values are reversed from the original, with whites appearing as blacks and blacks appearing as whites.

Nonunitized microform. A microform that contains unrelated information units.

Off-Line. A computer operating mode in which the equipment (described as being off-line) is not controlled by the computer system.

On-Line. A computer operating mode in which the equipment (described as being on-line) is controlled by the computer system.

Planetary Camera. A microfilm camera in which the film unit is suspended over a flat copyboard. The documents are placed on the copyboard and are photographed while remaining motionless.

Print. A duplicate of a master microform prepared on either paper or another microform.

Polarity. The tonal relationship between the original film or paper format and the copy made from it. A negative polarity is one in which the tonal values are opposite; a positive polarity is one in which they are the same.

Positive Image. A photographic image in which the tonal values are the same as the original's, with whites appearing as whites, and blacks as blacks.

Pull-Down. The distance between individual microimages as they appear on a microform.

Reader. Equipment that is capable of enlarging microimages to a size that can be read with the naked eye.

Reader-Printer. Equipment that in addition to enlarging microimages to a readable size can also make enlarged paper copies of selected microimages.

Reduction Ratio. The ratio that indicates the number of times that the original document has been reduced while being converted to microfilm.

Rejuvenation. A process that removes superficial scratches and stains on microfilm that has been coated with lacquer. This lacquer coating is removed with solvents and a new lacquer coating is applied.

Resolution. The ability of a lens or photosensitive material to separate closely spaced information on a microimage. Resolution is expressed as lines per millimeter.

Roll-to-Roll Duplicating. A contact printing process that will reproduce the microimages contained on one continuous length of microfilm in a second one.

Rotary Camera. A microfilm camera in which the documents are transported by a moving belt before the film where they are photographed while still in motion.

Silver Halide Negative Film. A photographic film consisting of silver halide crystals suspended in a gelatin emulsion that releases free silver upon exposure to light and a developing agent.

Silver Recovery. The process of reclaiming the free silver contained in developing solution and the silver remaining in exposed film.

Source Document. A document that provides information required for decision making or the further processing of data. *Also:* the document that serves as input to a microfilm camera.

Step-and-Repeat. A photographic technique in which a group of documents is exposed in multiple rows upon a sheet of film. The camera automatically positions the microimages on the film.

Targets. Markers filmed as part of a microform that fulfill one or more of the following purposes: determine resolution, identify the microimages; sectionalize the roll microfilm for ease and speed of retrieval, certify the microimages as true copies, and so forth.

Trailer. A supplemental microform containing microimages relating to a previously prepared microform.

UFC. (Acronym). Uniform Photographic Copies of Business Records as Evidence Act.

Ultrafiche. Microfiche in which the reduction ratio exceeds 90×.

Uniform Photographic Copies of Business Records as Evidence Act. Legislation passed by the U.S. Congress in 1951 that provides for the acceptance of certified microfilm copies of business records as primary evidence in federal courts of law and in legal procedings. This title, and its acronym (UFC), is generally also used to describe the various States' versions of this Act.

Unitized Microform. Microforms that are planned as one complete unit or subdivision of information without reference or attachment to any unrelated or extraneous microimages or microforms.

Vesicular Film. A type of film in which the image is formed by the application of heat and ultraviolet light.

INDEX